THE
CENTER
OF LIFE

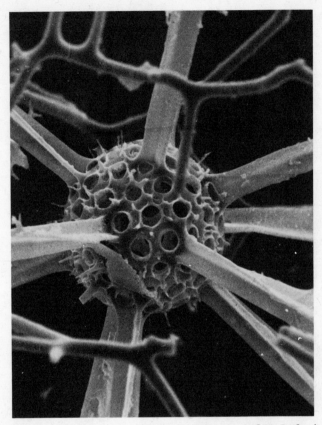

(courtesy of J. and M. Cachon)

Plate I. At the center of *Lychnosphaera regina*, one of the elegant amoeba architects. "I have the suspicion that we're not the innovators we think we are, we're merely the repeaters. It took us an embarrassing three to five million years to develop the architecture the amoebas have had for a few billion."

THE CENTER OF LIFE

A Natural History
of the Cell

L. L. LARISON CUDMORE

QUADRANGLE BOOKS
Published by Quadrangle/
The New York Times Book Co.

First Quadrangle paperback edition, 1978

Quadrangle Books are published by Quadrangle/The
New York Times Book Co., Inc., Three Park Avenue,
New York, N.Y. 10016.

Library of Congress Cataloging in Publication Data

Cudmore, L. L. Larison
    The center of life.

    1. Cytology.   I. Title [DNLM:
    1. Cytology—History.   2. Cells.
    QH581.2 C964w]
    QH581.2C83 1977   574.8′7   76-50822
    ISBN 0-8129-6293-1

For the men of my life:
Patrick, Colin, and Sean

The author would like to acknowledge, gratefully,
the kind collaboration of Miss Rosamund Gifford.

Contents

CONTENTS

CONTENTS

It has been customarily said by the public journals, assumedly bespeaking public opinion, that scientists "wrest order out of chaos." But the scientists who have made the great discoveries have been trying their best to tell the public that, as scientists, they have never found chaos to be anything other than the superficial confusion of innately a priori human ignorance at birth—an ignorance that is often burdened by the biases of others to remain gropingly unenlightened throughout its life. What the scientists have always found by physical experiment was an a priori orderliness of nature, or Universe always operating at an elegance level that made the discovering scientist's working hypotheses seem crude by comparison. The discovered reality made the scientists' exploratory work seem relatively disorderly.

—R. Buckminster Fuller

THE
CENTER
OF LIFE

ONE

◆━◆━━◆━◆

The Universal Cell

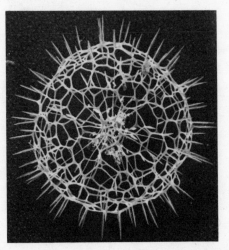

Plate II. *Actinosphaera capillaceum.* "*Our* cells, the ones we love, are repositories of such fantastic architectural flights—pleasure domes far beyond even the most opiated dreams of Coleridge...."

All cell biologists are condemned to suffer from an incurable secret sorrow: the size of the objects of their passion. Almost anyone with an obsession can share it with someone else. A numismatist easily lifts his gold, silver, or bronze inamorata from felt-lined mahogany drawers. Even an astronomer needs but a clear night to display his prized and flaming gems. But those of us enamored of the cell must resign ourselves to the perverse, lonely fascination of a human being for things invisible to the naked human eye. Never can we easily extend the invitation, "Let us go then, you and I . . ."

Some cells are extremely visible—the egg of an ostrich, of a hen or puffin. But we cell biologists see these the way anyone would, as a large globe of yellow yolk surrounded by a transparent glutinous mass; interesting only by virtue of their behavior in soufflé or omelet. We are happy to leave those particular cells to medieval painters or any others who love the transparency and staying power of egg tempera. For us a hen's egg is a delicious, marvelously packaged, but essentially boring cell, with no resemblance to those miniature baroque houses we have visited, the ones so sterilely and forbiddingly dubbed "the basic unit of life." *Our* cells—the ones we love—are repositories of such fantastic architectural flights; pleasure domes beyond even the most opiated dream of Coleridge, a Xanadu percolating with the directed chaos of those hundreds of thousands of simultaneous chemical reactions that are life.

This universe we love is a distant one and a hard one to share. Microscopes are damnably difficult to learn to use and something not commonly found around the house. Cells have to be killed, sliced, filled, with plastic and exotic metals before their intimate parts can be photographed, and even then, for most people, the photographs end up being meaningless collections of blobs and lines in black, white, and gray. Even the vocabulary is prohibitive; esoteric and sometimes poetic but about as familiar as Uzbek or Urdu. Mitochondria, flagella, cilia, nucleolus, endoplasmic reticulum. The parts of a cell make a litany of near-gibberish, the only one with an honest American name is the nucleus. We are a sad lot, the cell biologists; like the furtive collectors of stolen art, we are forced to be lonely admirers of spectacular architecture, exquisite symmetry, drama of violence and death, nobility self-sacrifice and, yes, rococo sex. All found in the world of the cell. Cells have everything. But visibility.

Every living thing is made of cells, and everything a living thing does is done by the cells that make it up. That is the truth and there are no exceptions; one reliable, unchanging fact in a changing world twenty-seven years after midcentury. These dictatorial cells have total control, even of our bodies, those of such a superior creature as *Homo sapiens sapiens* the wisest of the wise, a sentient, cognitive creature of supposedly independent will. Yet we don't make a move but that some group of the sixty thousand billion cells in our body makes its move first. Cells let us walk, talk, think, make love, and realize the bath water is getting cold. Cells that can translate light into electrical impulses let us see where the taps are; then we need more cells to turn on more hot water—nerve cells, muscle cells, and a biochemistry of which only cells are capable.

Unbelievable, unsettling and sometimes unexplained things are going on inside our bodies. And the beginning of all of this is a single cell, a fertilized egg. Frantically swimming sperm meets lonely awaiting egg. They fuse. The egg gives a small, swift shiver, a slight

6

electrical *frisson*. A wave of electrical charge sweeps over its surface. A tiny thunderbolt and *shazam!* a new organism begins. All that organism can ever be is no more than can be carried in this pinpoint-sized fragment of jelly. This is spectacular: the start of everything is simply this single cell splitting in half and then in half again. And again. One cell becomes two cells, then four cells, eight, and ultimately two thousand billion cells. No longer just cells, but a child with hair, fingernails, a heart, hair, and stomach. This isn't solved yet, how a single cell gets to be such a wonderful thing as a baby mammal. Nor do we know how from that mere cell beginning we get both red blood cells and nerve cells. Cells firmly believe that form should follow function, and one fertilized egg gives us the blood cell, a plump, thin-walled jelly donut of a cell, a sack of hemoglobin pushed around the body from lungs to heart and back again, and the flat star-shaped nerve cells, radiating as many as 60,000 electronic connectors. They look different and they do different things: the red blood cells run their course, endlessly delivering oxygen and taking away carbon dioxide, the waste product of life, to dump it in the lungs; the nerve cells are delicate transmitters and filters in a complex electronic circuit. But nerves and blood cells are only superficially different. Nerve cells, red blood cells, skin cells, or sperm, like the colonel's lady and Judy O'Grady are sisters under all their differing skins. They share the same structures and the same chemistry.

It's not quite as simple as saying, "See one cell and you've seen them all." Actually you have to see *two* cells to have seen them all. The world is divided into haves and have-nots even in celldom. All living things are made up of either of just two kinds of cells: those that have a nucleus and those that haven't. Everyone—

haves and have-nots—carries their hereditary information in DNA: two extremely long strands of it, wrapped
around each other in a double helix. This is the re-
pository of all of the knowledge a cell has; an encyclo-
pedia and guide of what to do in all situations, possible
and impossible; which molecules to make and use,
which conditions to seek, which to avoid, how to move
. . . how to survive. All living things need their instruction manual (even nonliving things like viruses) and
that is all they need, carried in one very small suitcase.
If they needed to, twenty-five furtive cells could hide
under this period. The blueprints for the construction
of one human being requires only a meter of DNA and
one tiny cell. That's all. It's comforting to know that
even Mozart started out this way.

Everything has DNA, but not everything wraps it up
in a nucleus. As might be expected, the have-nots are
less careful about their possessions; their DNA is out
in the cell, rubbing shoulders with the less vital, less
elite structures. It's more democratic, perhaps, but has
been a hindrance to evolutionary advancement. The
have-nots (prokaryotes) are not organisms we usually
associate with truth and beauty—bacteria, slime molds,
and the algae choking stagnant ponds are some—but
they are amazingly resourceful. We can perhaps admire them because they are the least fussy and most
adaptable of us all. However, I doubt if they need our
admiration for they are truly successful; they are found
universally in every nook and cranny of this world, including those nooks and crannies where nothing else
could stand it. If we travel to the most unpalatable,
most Godforsaken and least appealing place on our
planet earth, there we will find some prokaryote living
in great contentment. The lower reaches of our colon,
the excruciating cold of a bleak Antarctica, the sul-

furous hot springs of Yellowstone, and the depths of the airless muck in a sweltering swamp . . . there we will find the prokaryotes. They also live everywhere else. We haven't found them outside of our planet, but we're looking and we're expecting them, especially on our blushing neighbor, Mars. Mars is cold, barren, and a terrible neighborhood to live in. No doubt some prokaryotes have found it comfortable. Their needs are simple, their wants few, and their nutritional ingenuity outclassing that of even the most successful graduates of survival training.

Prokaryotes can always find food and a home, but they still haven't made it. They are all only single cells, and their lack of a nucleus has kept them in their lowly state and will keep them there forever. They can never become a multicellular organism. Denied to them forever is the chance to become something great and awe-inspiring, like an insect or a human being. Only the haves—the ones with nuclei in their cells can make up the bodies of the marvels of creation—those with hearts, lungs, kidneys, and minds. To have any of these exceptional organs, you first have to have the capacity to form a tissue. And versatile as they may be, prokaryotes cannot do this. No prokaryotes have ever been known to come together in an organized congregation of cells that look the same and are doing the same thing. And if you can't make a tissue, you can never make an organ, whose functions are truly awe-inspiring.

The kidney is not what you would call a popular organ, except as a culinary treat for some, and it certainly has not inspired romantic feelings in any of us, as the heart and the brain often do. But the talents of the kidney are truly prodigious. No larger than an honest man's fist, it can make any chemical engineer blush at his ineptitude. It handles waste disposal, recycling, and reclamation with ease. It can be replaced only by an unwieldly mass of wires, electricity, and tubes. And we get two of them—without charge—at birth. It is the cells in the kidney that are responsible for performing the interminable tasks of filtration and concentration, and end up tasting good, too.

We all have some idea of what our minds do. We've got a truly marvelous brain, where consciousness, associations and volition, and perhaps our "humanity" reside. IBM's best cannot touch those two "handfuls of porridge" under their bony arches for sheer complexity, computational talents, and information storage and retrieval. Not to mention the fingers, eyes, ears, and tongue that are electrically coordinated extensions of this mushlike computer. If it could build one, IBM would charge millions of dollars, but we get our brain free, courtesy of cells and evolution. Yes, we can trace consciousness and volition to the brain, but how much of that volition is subject to the cells that make up the brain? It's something to think about: whether we have a will of our own, or whether our will is that of the tiny collections of molecules and biochemistry, the cells.

Well, if it is, we could do worse. Almost all the time cells subjugate themselves to a rigid code of social behavior. They follow rules and are receptive to signals and requests from one another. The rules govern both growth and territory, and they have to be followed perfectly. Otherwise we wouldn't end up divided into fingers and toes, eyes and brain; we would end up a hopeless and shapeless jumble.

But cells are normally very social and well behaved. They will stop growing and dividing when they come into contact with another cell. And as members of society, not even the early Christians were more giving, more humble, less self-seeking, and more dedicated to a single thought: the survival of their organism. Of course there are renegades, cellular Charles Stark-weathers and Richard Specks, cells perhaps more frightening to us than even their human counterparts. I don't know how many people are more frightened of being done in by a mass murderer than of any other fate, but 65 percent of Americans are more frightened of cancer than of any other disease. Cancerous cells are crazy; antisocial cells that won't obey the rules and won't stop growing, grabbing more than their share of space, blood, and food. Traitor cells these, that threaten their fellow cells and the entire organism by starving out and smothering normal cells. One small brain tumor, and all of our sophisticated organic equipment goes for naught.

If I were a nineteenth-century writer believing in the inherent goodness and perfectibility of mankind, I would no doubt urge my readers to emulate in their own societies the faultless behavior of the cell in the society of life. Cells are obedient and noble, perpetually hard-working, devoted to the health and survival of the organism they form. I'm not a nineteenth-century writer, and almost all my Victorian belief in the innate goodness of human beings has been erased by Buchen-wald, Siberia, Vietnam, and the Turkish slaughter of millions of Armenians. I don't think it would do any good to urge people to act like cells any more than it has done us any good to have the industry of ants and the altruism of honeybees pointed out. We aren't honey-bees or ants; we are human beings. They went their way 500 million years ago and we continue on ours, probably not interested in being good and obviously not given to learning moral lessons. Maybe that's why I like cells so much.

We are basically the prisoners of our cells, and I find

it pleasingly paradoxical (if somewhat demeaning) to be in the thrall of something that small and mindless. We can be no more than our cells together are. At least our cells are the haves, the eukaryotes with nuclei. It was decided about two billion years ago. This advanced stage of evolution—each cell a true member of society—is available only to the eukaryotes. The prokaryotes can never belong, they are doomed to perpetual single cellularity, at the mercy of the haves, to be eaten by them or destroyed by their antiseptics and antibiotics. Prokaryotes can never hope to become a kidney or even a part of a kidney. Their discontent may be great, for their revenge upon the eukaryotes is very great indeed. Think of a terrible human disease—syphilis, cholera, the Black Death, leprosy, blood poisoning, gonorrhea—the Prokaryotes' Revenge. Bacterial diseases all.

Life has been going on for at least three and a half billion years, beginning as something very like a single prokaryotic cell. Since then it has been shaped and sculpted over and over, into roses and giraffes and fleas, but it seems that one general principle has always been followed. The evolution of cell societies parallels that of human societies: there has been a constant movement toward specialization of the members of the society. Increasingly, totally competent and independent individuals have become associated into complicated collections of specialized members. In the beginning, one member could do everything: make weapons, catch food, digest it, get rid of wastes, move around, build houses, engage in sexual activities straightforward or bizarre. Once they enter their societies, the members—whether cell or person—tend to become specialized, performing only one of the many tasks performed by the whole of which they are part (all of course retaining the right to reproduce).

THE UNIVERSAL CELL

This has been a gradual change; we still have examples of the beginning and intermediate social experiments. Jellyfish and their relatives, especially the lovely but dangerous Portuguese man-of-war. These are just loose cooperative associations of cells, each Portuguese man-of-war—an inflated plastic bag hung with confetti streamers—is not just a single individual. It is really a colony of hundreds of individuals divided into four classes or trade guilds; these are those that make up the float; those that form the long feeding tentacles; another set that are the defenders of the colony, forming the poisonous, sometimes paralytic stinging cells; and finally those that take care of all of the reproduction for this floating commune. Although there is such a division of labor, all individuals can take over any of the tasks. If the float is destroyed, any organisms making up the feeding tentacles or the defense system can take over and make a float.

With further evolution, we became more and more specialized; now an organism needs thousands or millions of cells to do all it has to do, and each cell has little independence. It can perform only one task and cannot survive outside of the total organism. Nor can it adapt to new tasks; a kidney cell can never—even in an emergency—take over as a heart cell or as a muscle cell. Similarly, in human societies, although each of us stubbornly retains the right to do his or her own reproducing, we don't need to kill our food, make weapons, or get rid of our wastes. The local supermarket, the Pentagon, and the department of sanitation are there instead. Not too many of us could do all of the necessary things on our own. At one time we could. And at one time, about 600 million years ago, just a single cell could make weapons, catch food, digest it, get rid of wastes, move around, build houses, engage in sexual activity straightforward or bizarre. These creatures are still around. The protists—organisms complete and entire, yet made up of just a single cell with many talents, but with no tissues, no organs, no hearts and no minds—really have everything we've got.

It is safe to say that, as a group, the protists do not form an integral part of the American consciousness. The amoeba is the only one that has achieved much media recognition. Its name is used as a pejorative synonym for anyone without personality, suspected of leading a spineless, creeping kind of existence; and the threat of amoebic dysentery keeps travelers wary of the local water supply. We are both slandering the amoeba and missing a lot, including some very poetic names: *Chaos chaos* (with a name like that, it's *bound* to be interesting); *Acinetopsis rara; Ephilota gemmipara; Gonyaulax polyhedra; Noctiluca; Stentor ceruleus.* The names don't exactly spill trippingly off our tongues like the syllables of Lolita off the tongue of Humbert Humbert; but they are beautiful, and so are the namesakes. Not as beautiful as some nymphets, perhaps, but what nymphet can glow with a cold mysterious blue-green fire like *Noctiluca,* igniting Pacific or Atlantic waves into glorious fireworks? *Noctiluca* (nightlight) is luminescent, lighting up the marine night, the source of phosphorescence in the ocean. *Stentor,* after the trumpet-voiced herald in the *Iliad,* is a large vivid-blue trumpet, with waving fringes around its bell. Though attached by a stalk to any surface, if it is annoyed by taps of a pencil, it will first contract; just as anyone else would hunch his shoulders and pull his neck down into them. If repeatedly annoyed, *Stentor* will loose its hold and indignantly swim away; rather complicated behavior for a single cell with no nerves and no brain. But even the blue trumpets of *Stentor* are not the most

14

interesting denizens of this wild and beautiful Lilliput world, half Disney, half Dali.

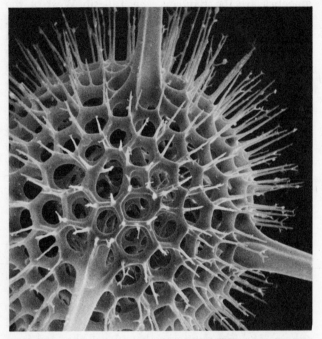

(courtesy of J. and M. Cachon)

Plate III. . . . it has to be three lacelike fretted glass domes, one inside another like the ivory follies that are the pride of Oriental ivory carvers. Magnified 950 times.

Ah, the architecture of this world. Amoebas may not have backbones, brains, automobiles, plastic, television, Valium or any other of the blessings of a technologically advanced civilization; but their architecture is two billion years ahead of its time. The amoeba had the architectural ideas of R. Buckminster Fuller before there was anyone around capable of having an idea. The amoebas are the Bauhaus, the Taliesin West of the protist world; though apparently Frank Lloyd Wright never talked much to them. He bitterly complained that he was the only architect who knew how to use glass; his fellow architects used it only to cover holes in their walls, ignoring the properties of glass as a structural material. But the amoebas ancient or contemporary are not guilty; some of them have built their houses completely of glass. Among these are those that are artisans, not architects, who merely secrete a sticky coat they cover with sand grains, using their oozing jellylike feet. Certain amoebas place these grains helter-skelter, but careful others match the edges of the grains perfectly, rivaling the precision of a Byzantine mosaicist at the court of Justinian and Theodora.

Other amoebas with a bit more architectural acumen make their homes of overlapping hexagonal glass plates, ending up as tiny, perfect crystalline pine cones or pineapples. The Radiolarian amoebas are rather flamboyant; they build glass sunbursts, with long thin transparent spikelets radiating from a central crystal sphere. The Foraminiferans have been liberated from such pure geometry; with only their boneless, formless feet, they make intricate constructions by adding chamber after chamber, ending up like a somewhat free-form chambered nautilus or like a Turkish turban. (Build thee more stately mansions, o my cell.) Finally we have amoebic geodesic domes to make Bucky Fuller weep with delight. (He already knows about them.) Glass struts are built into hexagons and used to make simple geodesic domes or fancier homes. *Cyrtocalpis urceolis* makes its dome in the shape of an urn; narrow at the bottom, gently swelling at the middle, and narrowing again at the mouth. Other geodesics are put together like onion domes stolen from (by) the Kremlin. And, as usual in any art, there has to be an overachiever. *Il sorpasso.* Our amoebic Beethoven, our gelatinous Gaudi is *Hexacanthion astercanthion.* One geodesic dome will not do for this superarchitect; it has to be three lacelike fretted glass domes, one inside another, like the follies that are the pride of Oriental ivory carvers; detailed hollow spheres nested within each other, the openwork of one revealing several more exquisitely carved balls.

There are different schools of architecture even down among the protists. Not everyone has chosen glass; some prefer the properties of metal—strontium sulfate. These usually construct twenty spines, all radiating outward from the center of the cell, and always arranged in perfect Euclidean geometry. The metal sculptures can be rather nice, but they do lack the drama of glass geodesic domes, especially a triple concentric set of domes which, by the way, is tastefully decorated with delicate projecting spines. And all of these are built by the much-slandered, unappreciated amoeba; a creature of very little brain, I can assure you.

Yet, for brainless creatures, the protists are both ingenious and ferocious when it comes to dinner. Many just ramble around their ponds inconsequentially, and if they bump into something edible, all well and good, they'll eat it. Amoebas will just ooze around whatever delectable piece of detritus they may meet, their eating habits far less fastidious than their housing. Other protists just randomly sweep currents of water into their mouths and gullets, hoping that something good will come in with the water; but then there are the dedicated, predatory carnivores with cunning inescapable

17

weapons and tactics. An entire group of protists are dedicated to meat-eating, prey-seeking pursuits. *Didinium* is one. Under the microscope it seems harmless enough, a rounded cup, bouncing aimlessly and gaily back and forth, up and down. If one should bounce into something good to eat, like a *Paramecium*—a hairy protist sausage, shaped like a slipper—we see a completely different *Didinium*. Immediately it sends out short rods from around its mouth and attaches itself firmly to its victim. The prey is then gradually paralyzed by the *Didinium*'s secret weapons, the infamous, deadly poison pexicysts! *Didinium* needs something, for *Paramecia* are at least three times as large and very strong swimmers, but no match for the paralytic poison of the pexicysts. The *Paramecium* quickly stiffens and ceases to move; then, without further ado, the whole *Paramecium* is crammed (really pulled) into *Didinium*'s mouth to be digested. *Didinium* expands to fit its prey; and then, disgustingly bloated, it wobbles off, lurching slightly, no longer capable of a spectacular bounce.

Poison is common among these carnivores, innocent Borgias all. Many have barbed poisonous harpoons that they can launch in split seconds. Not bad for something without a nerve to its name. Certain amoebas are carnivores also, but as predators they have a problem: locomotion. Oozing from one place to another can hardly be considered rapid transit. They manage to get around, but it is not exactly a hunter's pace. Any live prey could outdistance them as easily as a rabbit could escape from a carnivorous turtle, which is no doubt why turtles stick to worms.

Amoebas, however, manage to get their meals. It isn't clear how—poison, amoebic charisma, or hypnotism—but somehow the plodding amoeba is able to stop a healthy, rapidly swimming *Paramecium* in its tracks. A *Paramecium* will stand—like a faithful, well-trained nag—patiently awaiting a doom that is grue-some and no sight for the squeamish. Motionlessly it stands, while the amoeba oozes around it, finally en-gulfing it completely in its transparent shifting proto-plasm. There follows a scenario of which Poe would have been proud. Once the *Paramecium* is completely inside the amoeba, the spell is broken; it may be seen there, surrounded by amoebic protoplasm, fully alive, frantically beating the hairlike cilia covering its body (no doubt the *Paramecium* version of kicking and screaming) as it is slowly digested. We can read an-other version in *The Cask of Amontillado*.

Some protists don't even have the limited potential for movement that amoebas do; they remain perma-nently attached to a single surface. But they still have to eat. *Acinetopsis rara* grow on slender stalks, immo-bile, but they will sway on their stems like a garden of pale, translucent tulips. It is a deceptively sweet flower, blooming on its stalk, a true *fleur du mal* if there ever was one. For *A. rara* has a deadly device, a lariat. And so it sits, our palindromic *A. rara*, like a cowhand rest-ing on a corral fence, idly tossing its lasso, a long, thin, prehensile tentacle, out and back. Out again and back.

Our *A. rara* is quite a gourmet. Nothing will satisfy its appetite but one special dish, one kind of protist out of the more than 20,000 available. Only the rhyming *Ephilota gemmipara* will stay the hunger pangs of *A. rara*. Only if the interminably tossed lariat encoun-ters an *Ephilota* will anything happen, but then the tentacle attaches firmly and a fateful tug of war begins; for *Ephilota* also is firmly attached to its stalk. Eventu-ally there is a final tug; *Ephilota* is torn cruelly from its stalk and devoured. Sometimes uneaten pieces of *Ephilota* will still be clinging to *A. rara*, like grisly gobbets stuck in the beard of a slovenly gourmand.

Yes, everything was there 600 million years ago. The single cells did all right. With or without a nucleus, they did all right, for they have endured. And with or without a nucleus, they discovered the benefits of sex way back then. Now, sex is important. Aside from its recreational and entertainment possibilities, it has considerable biological significance. The biological significance was there first; the entertainment value came only recently.

Sex is really just a three-letter word for the exchange of genetic material. Many organisms—even those made up of many cells—can reproduce without sex: aphids, worms, honeybees, prokaryotes like the bacteria. Now this is fine if you're lonely, but it isn't fine for evolution. As flowers need the rain, evolution needs genetic variety. It has to have ever-new combinations of genes, new genetic experiments, one of which might work better than others have worked up until then. A new amoeba is born because an amoeba splits in half. This new amoeba is the spitting image of its parent. It has no choice. It has exactly the same genetic information to work with. This is impressive precision, but it is evolutionarily dull. The amoebas stayed where they were exactly because they didn't have sex. Without these genetic experiments where two parents combine their genetic information, we would have stayed amoebas, too; and *that*, oh best beloved, is why sex had to rear its ugly but seductive head.

We know sex is a good thing; even the bacteria practice it. One bacterium meets another and, if the genes are right, there will be a one-way exchange of genetic material. There are no eggs and sperm; in fact no males and females. Whatever bacterial mates are to each other, to us (through chauvinism or scientific precision) they are a coldly clinical "donor" and "recipient." Admittedly bacterial sex is a pretty coldly clinical process; the donor simply extends a tiny tube over to the recipient, and then unreels its DNA into the recipient through the tube. It took the protists with their nuclei to devise some plain and fancy variations on the theme. If no one interesting is around, a protist can just split in half if it has to; but some kind of creative imagination must have come along with the nucleus, for the protists have sex as elaborate and as varied as their architecture.

In the simpler cases, a single protist can produce male and female sex cells that, like sperm and egg, fuse to form a new organism. Even when they do engage in sexual reproduction, protists are never referred to as male or female. A protist can only belong to the more prudent (or prudish) "mating types." But sex is what it is that they do. Two individuals of opposite mating type get together and exchange genetic material, and some of them do it with great abandon and much imagination. As a group, the protists are devoted to sexuality. Even though many protists are just straightforwardly and pragmatically heterosexual—a pair just simply fusing their nuclei—not all are satisfied with this method. Some mate for as long as four or five days, taking the biblical invocation to become flesh of each other's flesh literally. Normal cell boundaries break down, and they actually fuse together.

Others have a basically uncomplicated heterosexual mating, but they require a crowd. You will find mating pairs only if hundreds of cells of opposite mating types have been mixed together. All of these cells stick together for a while forming a massive clump, and only later, as if reassured that "everyone's doing it" will one cell pair with another and swim off to fuse their nuclei.

The first protist I ever met was very demanding sexually. These *Paramecium* not only demanded one of these mass reactions, but they also had to be hungry, but not starving, and preferred to make their assignations between three and four in the afternoon. Really. This isn't what you could call group sex, although that does occur in one species. As many as fourteen individuals will get together in a group and mutually fuse their nuclei. While they are doing this, they present a truly charming spectacle; for each cell wears a small twisted glass turban.

Perhaps the most appealing aspect of protist sex is that they can have more than two sexes. Desmond Morris claims that we are the sexiest of all animals. Perhaps it is true; no doubt we would like to think it is. But all we can manage is heterosexual, homosexual, bisexual, or asexual. Consider the possibilities for those protists that have four, eight, or more mating types. They actually practice the simple unadorned heterosexuality with which we are so familiar; mating is by pairs, one pair at a time, each partner of an opposite mating type. If we were a protist belonging to a species with four mating types—A, B, C, and D—each of us would be just one of the mating types, and we could mate with all of the other mating types except our own. If A, we could choose B, C, or D as a sexual partner, but not A. That would be homosexuality, and it is about the only variation that isn't normally seen. Protists don't have any moral strictures, but mating types are determined by the interaction of molecules on their surfaces, and two cells of the same mating type just won't stick together long enough for anything to happen.

The advantage of an abundance of different mating

types is obvious, although we cannot consider boredom and the need for variety a part of protist existence. Having more than two mating types simply makes it easier to find someone to mate with. If there are only two sexes, and half the population is made up of the opposite sex, each individual has a fifty-fifty chance of meeting someone he or she can potentially reproduce with. As we know from experience, these odds aren't as good as they look on paper. If there were four sexes, we would be better off, having a 75 percent chance of finding an appealing member of the opposite sex. *Stylonichia* has the best odds of all; there are forty-eight mating types which could mean that each *Stylonichia* has ninety-seven percent of the population from which to choose a sexual partner. We can well envy *Stylonichia*— with those odds how could anyone ever be lonely? Yet considering our record of success with just the two sexes we are born into (and the few variations we might add onto these), I don't think we could handle it.

There is even a protist that engages in mating play. After two cells have chosen each other, they do a little dance before they mate. Lowering their mouths, they move jerkily around each other in small circles, like Indians doing a rain dance. Then they mate. I'm not sure I like this; I know I don't understand it. Why do they need to do it? A single cell can have everything except fire, and intellectually I like that thought a lot. Emotionally, I feel that cells have done too much. They seem to have accomplished everything we pride ourselves on, but they did it two billion years ago. Granted, their pexicysts are not exactly SAMs or ICBMs, but they're not bad at all for something without a brain and without hands, and they do what is needed without contaminating the environment. I have the suspicion that we're not the innovators we think we are; we're merely

23

the repeaters. It took us an embarrassing three to five million years to develop the architecture the amoebas have had for a few billion. Some terrific achievement.

Worse, and even more humiliating is that the protists have also come up with the idea of an eye. One stalked creature has built itself a darkly pigmented retina, equipped with a real lens. And there it sits on its stalk, a Cyclopean eyeball relaxing in an Eames pedestal chair. Seeing. Watching. That I really don't like.

TWO

Biochemical Evolution

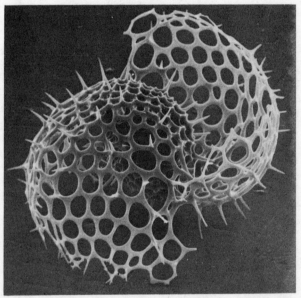

Plate IV. The spiraled helical skeleton of *G. lithelissae*.
". . . the exquisite dymaxion constructs of mindless
amoeba." Magnified 470 times.

W e are made of cells. And of stars. The universe outside of us has made the universe inside us. Those atoms created in fusion and fission at the beginning of it all went their ways and scattered themselves through the universe. Some ended up in that large molten ball careening through space in a complexity of spins—our planet earth circling on its own axis and around a star that makes its own spiral through a galaxy again whirling. In this Ptolemaic set of dizzying cycles and epicycles, protons became neutrons became electrons and then hydrogen and helium, then nitrogen, oxygen, and carbon; finally captured in the cooling mass a nascent life-supporting planet. Since atoms never die (well, hardly ever), those that were in at the creation—whether it was a big bang or just a smaller bang somewhere out in the suburbs of a constantly renewing steady-state universe—are still around. From the atoms that took up residence in a hot and gassy mass orbiting around a star of plus 4.8 magnitude, later cooling to become our home, from these atoms are we made and all things on the earth, those that crawl on the land, swim in the sea, and fly in the air, etc.

My best friend says that even back then, at the very start, these atoms had a consciousness, and that they still have it; complex things like memory are just macro-events that must be made of micro-events like molecules. He feels that gravity and mass attraction are just partial manifestations and good evidence of this. How else do you explain that, if there were only two hydrogen atoms in all of space, they would "know" how to find each other? This atomic consciousness directs the formation of increasingly complex states, so that atoms not only remember what has happened to them but also are directed to form molecules in certain ways. This consciousness may just be geometry, the deity of R. Buckminster Fuller "the extraordinary generalized principles and their complex interactions that are apparently employed in the governance of universal evolution."

So it is this universe geometry that winds the wonderful helix of DNA and planes the square clean-edged symmetry of a crystal of salt, repeated whenever salt exists. It also choreographs the precision waving of a plurality of spirochetes (those corkscrew-shaped bacteria that brought you syphilis and relapsing fever); whenever spirochetes get together, whether two or two hundred, they will thrash together in a spontaneous and sinuous unison that makes the Rockettes look like a centipede with coordination problems.

This consciousness also guides a pea tendril tentatively seeking support. Is it true that all of these grow or move, a crystal and a tendril, because of the memories and impulses of atoms that have made the molecules that made the tissues that become the organs that inexorably build the organism, the final and formal house that universe geometry has built? Does it all go back to what atoms learned as they were hurled through space? My best friend thinks so. I'm not sure. I don't think I can believe with my friend that atoms have memories but, on the other hand, he can make boats fly and projectiles that go in an absolutely straight line. He is often, though not always, maddeningly, presciently right about things. And it does explain those exquisite dymaxion constructs of a mindless amoeba. I am, however, fairly convinced of one thing: that a molecule has its properties by virtue of the atoms that make it up and that life has its properties by virtue of the molecules used in constructing living organisms.

So the structures of the cell were really determined

by what went on about six billion years ago while the earth was still cooling down. The structures of our cells, too. We were there before we even were. On the wet and primitive earth, floating like invisible noodles in the primeval soup, our molecules were engaged in the most intense conflict, ruthless competition for survival. It was a time of experimentation; vast, mindless experimentation. Molecules were being put together and broken down again and again. Millions of them must have been formed from the atoms that made up the chemistry of the primitive earth, but only a few have emerged triumphant. The list of survivors is short: nucleic acids, proteins, carbohydrates, fats, and lipids are the few that survived. The remarkably few, unbelievably few when we consider that every stitch, every pattern in the vast intricate tapestry of our biosphere required only these few organic molecules, the molecules of life. These molecules not only make life, they define it and characterize it. We don't find them where there is not life. They can be made in the laboratory, but only by a scientist who has to be very much alive.

Each of these kinds of molecules does something very, very well, better than any other molecules of six billion years ago when our biochemistry was decided. *Our* biochemistry. Although we are hopelessly *nouveau arrivistes* relative to most life on this planet, we are joined into a biochemical brotherhood with both a crude primitive like the shifting amoeba and a long-established aristocrat like an insect. Simply *everyone* got their biochemistry way back then. Everyone still has it for the same reason all Russian women once wore the same frowsy, baggy dress. Not because it was Soviet *chic* or because they wanted to, but because that is all there was. Very democratic (or socialistic perhaps).

But the biochemistry is just as tyrannical as our tyrant cells. Life really turns out to be a chess game. The rules are strict, allowing for certain individual strategies but unbending and unbreakable; the outcome of the game is determined not so much by chance as by the stringent rules that limit the possible moves. No parent—whether human parent, crabgrass parent, or bacteria parent rumbling around the lowest reaches of our guts—makes an offspring without using nucleic acid organized into genes. That way, human beings fortunately always produce babies, and crabgrass, unfortunately, always produces more crabgrass. The other molecules are equally universal. Everyone uses carbohydrates to get energy, and the great American disease, "fat," whose epidemic proportions are mirrored in the litter of reducing plans —books, exercises, or some diet devised by a thin but shifty-looking physician with an appropriately thin but dyed moustache—tells us how well molecular "fat" packs away the energy we have taken in but don't need.

Evidently there were no molecules more versatile than the proteins. Muscles, hair, teeth, fingernails, feathers, scales, gristle: proteins every one. Truly the Renaissance molecules are proteins, because they also have given us what no cell should or could ever be without: those catalysts who have earned their popular fame by tenderizing meats and removing stubborn stains and ground-in grease.

The enzymes. Enzymes make things happen. Madison Avenue has promised us that so many products will do that: a toothpaste, a sleek shining automobile, a whiskey, a shampoo. But Madison Avenue does not love the truth, and often we follow their advice only to be disappointed. Enzymes will never disappoint us; they always deliver. They do make things happen, causing crucial reactions and interactions. Nothing at all can happen in our cells without the catalytic presence of one or more enzymes. Oh, those enzymes. Once upon a midnight dreary, a student of mine (then an art history major, now a goldsmith) burst into my laboratory, breathless from running the mile or so from his dormitory room. While pondering weak and weary over his Natural Sciences 5, he had experienced a true James Joyce "epiphany"; he had the transcendant realization that, toiling away from him, unknown, unloved, and unsung, were thousands—hundreds of thousands, even—of enzymes. He was truly transported. Not everyone can easily achieve such a nirvanic state of enzymic appreciation. We might try. We have nothing to lose. Establishing a mystical oneness with our enzymes would no doubt expand our consciousness immeasurably. And they don't ask for money.

Few are the molecules that make us, and fewer still are the tasks that living things have to do in order to be alive. Aside from growing, feeling, and reproducing, whatever else we do, though attractive, interesting, or even fun, is needless frills and frippery. Actually, something nonliving like a crystal can grow, and into shapes and forms more beautiful than many living creatures. Emeralds growing and glowing in fiery green tetrahedra are more beautiful than a slug (to most people). So maybe all we really have to do to be alive is to be able to reproduce (you don't have to *do* it) and be able to feel. This "life"—though its definition is brief—takes energy. You see, life is order, death is disorder or entropy. To keep order from descending into entropy (chaos) requires a constant input of energy. This is why we eat. Plants make their own food (and so do some protists and prokaryotes). Give them some sun, some rain, a little carbon dioxide, and a few diverse salts, and like some terribly clever housekeeper who

31

can make seafood crepes from leftover tuna sandwiches or a velvet couch from wire coathangers and lint, plants will make all of the carbohydrates, proteins, and nucleic acids they need. They also have to make all we need. Since we are not clever like the plants, we have to eat them for the products of their cleverness or else eat animals that eat plants. So, it is a very good thing that we are all made of the same atoms and molecules; otherwise there wouldn't be anything to eat.

Plants cannot do a lot of things. They cannot play tennis, go for a walk, or scratch themselves, but they do have the one absolutely spectacular trick never found in the animal repertoire: they can eat the sun. They transform the energy of sunlight into carbohydrates and thus get all of the energy they need or want. Free and forever. Not only are we made of stars, but we end up eating a star, too. Funny how things come 'round. Whenever we eat a plant (or an animal that has eaten a plant), we link up to a long, direct chain beginning in the sun and ending in some movement, mental or physical, that we make.

> What the hammer? What the chain?
> In what furnace was thy brain?

Thy brain (and mine) were forged, and our fearful symmetry framed, while a wild scenario was being played out in the furnace of the cooling earth. The oldest rocks on earth are over six billion years old, and the oldest forms of life, turned to stone, are found in rocks nearly four billion years old. Somewhere in between, life began. The German Bible says, *"Und die Erde war wüst und leer."* "And the earth was waste and empty." With apologies to Martin Luther, it wasn't.

BIOCHEMICAL EVOLUTION

Chaotic and violent, but not empty. A dramatic though hardly inviting landscape this earth of ours presented in its early years. Hot and stormy seas squalled across its face. Volcanoes erupted. Lightning flashed constantly from skies that were filled as constantly with storm clouds. Water vapor, heated by the cooling earth and its volcanoes, would roil upward, forming magnificent thunderheads. When their watery burden became too heavy, the clouds would open releasing torrential rains and flashing Jovian bolts. Blake could not have asked for more appropriate "distant deeps or skies" for birthing tygers or life. There was no oxygen to breathe in this hellish place, even if there had been something around that wanted to breathe. If you felt like breathing at the time, there was a pretty choice: ammonia, carbon dioxide, or methane (the gas responsible for the fragrance of cow pastures).

If anyone had thought about it before 1924, they would have assumed that oxygen was present then and has been present, as it is now and ever shall be, throughout the history of the earth. But the Russian biochemist Oparin straightened us out on that in 1924. "No oxygen," he said. The earth was too hot for free gases like oxygen to have stayed around. They would have been driven off. Also, atoms have to be made into molecules for life to begin, and free oxygen would have destroyed the molecules we are familiar with, would have literally burned them up. We know that didn't happen. It couldn't have. We are here with our oxygen-sensitive molecules. So there was no free oxygen, but oxygen was there, bound into the water and carbon dioxide. It may have been *wüst*, but it wasn't sterile, our earth; all of the ingredients were there to make the primeval soup: a thick, warm organic broth, a rich molecular minestrone, full of methane, ammonia, nitrogen, carbon, and dissolved organic salts.

> Soup of the dawning,
> Beautiful soup.

And from this soup life was ladled out. In his book, Oparin wrote, "In the beginning . . . this is the way it was. It looks like this is all that was there." And all of the learned men disputed, from 1924 until 1953. Was there oxygen? Could life have arisen under these conditions? How many angels can dance on the head of a pin? On and on.

Then, in 1953, a "little child" came among them and ended their monkish disputations, writing a chapter in the history of biology that has always delighted me. The "little child" was a mere graduate student, Stanley Miller by name. Said Stanley (innocently), "Why not find out whether anything could have happened under those conditions?" No one seemed to have thought of that before as a possible way of ending the argument. ("How extremely stupid not to have thought of that," said Huxley when Darwin's theory was explained to him.) So Stanley Miller built an apparatus; truly one any of us could build in the kitchen or garage. A wonderfully simple Miller machine, one that counterfeited the primitive earth, and whose Pyrex bowls ended the disputation once and for all. It was just two Pyrex glass globes connected by tubing. An infernal machine that could replay the violent and inhospitable scenes of some billions of years ago. Miller mixed a simple but rich soup: water, methane, carbon dioxide, ammonia. He then heated his soup in the apparatus and sent it around and around and around. Each time the steam from the heated soup passed through a chamber where two electrodes passed a thin spark between them. A weak, almost pathetic mimicry of those huge, primordial thunderbolts. Around and around Miller's magical life percolator; from boiling bubbling soup to spark chamber and back again. For one week. Then, on the seventh day, Miller looked at what he had done and said, "It is good. It is more than good, it is terrific!" From simple, everyday inorganic molecules, Miller had created organic molecules, amino acids, the subunits of proteins; a molecule of "life" never found outside of a living organism. No fiendish manipulations of a Frankenstein were needed, just a touch of steam, a dash of spark, and a few mundane molecules. Life itself wasn't created there in Stanley's magnificent machine, but the material of life was. And, having shown them all how easy it was, Stanley Miller rested. Since 1953 dozens of organic molecules have been re-created this way; simple inorganic compounds sparked by a source of energy for putting the new molecules together. Copies of Miller's apparatus have yielded nucleic acids, proteins, and carbohydrates, but strangely these experiments were not done by Miller.

>An endless succession of tedious tasks
>Become hypothesis in Pyrex flasks.

I wrote those lines once, in a Prufrockian parody, long ago when I was young and in the first stages of a love affair—with science. I washed laboratory glassware, and I was learning what "really clean" really means. I was also learning what science is; that there is science and there is good science. Almost anyone can do science; almost no one can do good science. The first scientists

I ever knew were just ordinary human beings who did
science:

> And there will be time.
> Time to question, time to do.
> Time for bridge at 12:02.

Stanley Miller did some good science; those elegant
experiments that say, about something important, "This
is so. This is just so." And the nicest thing about it is
that good science is almost always so very simple. *After*
it has been done by someone else, of course.

Miller showed us where the molecular Big Four
could have come from. He led us to the edge of a giant
abyss, and the only ropes we can throw across at present are those of thought and dream, not experiment.
The question we want to answer now is how we can
get these molecules to live and make a cell out of
them. It happened once; we know that at least. But as
Oparin has said, we can have about as much of an
idea of what went on so many billions of years ago as
a jellyfish can have of how life is on land. There may
be a masochistic streak in some scientists (or to be
more kind the same streak, whatever it was, that drove
men after questing beasts or Holy Grails), for it is a
cold, hard fact that we will never find out for sure how
the first cell arrived on the scene. Knowing that we can
never know doesn't seem to stop them from trying to
figure it all out, though they don't have a jellyfish's
chance.

Everyone is proceeding on at least one major assumption: that Darwin was right. There always has existed
a ruthless force deciding which forms live and which
die. This force is natural selection, and it does not suffer
fools at all gladly. Only the best competitors in the

struggle for existence facing every living organism win the medals. In this case, they are allowed to live. Evolution is a hard, inescapable mistress. There is just no room for compassion or good sportsmanship. Too many organisms are born, so, quite simply, a lot of them are going to have to die because there isn't enough food and space to go around. You can be beautiful, fast, strong, but it might not matter. The only thing that does matter is whether you leave more children carrying your genes than the next person leaves. It's true whether you're a prince, a frog, or an American elm. Evolution is a future phenomenon. Are your genes going to be in the next generation? That is all that counts. Anything else is, as my father used to say inelegantly but expressively, as useless as a bosom on a boar hog. So says Darwin's theory. It's still a theory. It fits most of the facts, but that could be because it's just consistent with our jellyfish view of the world.

Before we attack the problem of turning aimless molecules into a living, breathing, reproducing cell, we are going to have to gird our mental loins with a lot of faith and with another theory. I'm sure you all will recognize the ever-popular principle of competitive exclusion. It is simple, straightforward, and somewhat frightening: two different organisms cannot exist in the same space at the same time if they are competing for the same materials. One will eventually be excluded (= dead). We think that this applied to molecules as well as to living things. It accounts for the survival of the Big Four and the loss of all of the others, out of the many millions of molecules made as atom jostled atom in the primal soup. Any molecule that happened upon the atomic secret of survival—how to stay together, grow, make replicas, and use less energy in doing this than another molecule used—won survival (and perhaps had been coached by a higher atomic consciousness). Out of the millions of experimental model molecules, only the Big Four survived, and their rule over our biochemistry has endured ever since then. A reign that makes the dreams of the planners of the Thousand Year Reich as pathetically paltry as their souls. (Well, perhaps not quite *that* paltry.)

I hesitate to mention the outstanding feature of life so soon after mentioning the Third Reich. It may invite the reader to make an invidious comparison, maligning our benevolent dictators, the innocent cell. Life simply has a passion for order. In the only unit of life we know, the cell, everything is ordered; there is carefully controlled direction. What goes on within the cell is different from what goes on without the cell. Order defines the cell as it defines life. Before there was life there had to be system. Chaos may be creative, but it just won't do for every day. There has to be order. A bit boring, perhaps, but it's life. Whence this tendency toward tedium, this never-ending cellular devotion to predictability? We don't have to look to anything grander or more alive than a drop of oil. We have seen such a system thousands of times: in salad dressing. Oil and vinegar don't mix. Nothing can make the golden globelets of olive oil merge their identity with that of the vinegar. We can fracture the oil drops, but stubbornly they re-form, coalescing in refractive globules shining with light. This latter property is lovely, but not evolutionarily crucial to our story. The former property is. What keeps oil and vinegar (or oil and water apart) is not sheer perversity or prejudice, but the unlikely story that oils behave like male musk oxen, and that like musk oxen, they have heads and tails: water-loving (hydrophilic) heads and water-fearing (hydrophobic) tails. Whenever certain kinds of fats get together, they act just like male musk oxen, protecting their wives and children from wolves. They circle together, heads outward, tails inward, side by side. This is just something that certain molecules do. It is a good strategy for musk oxen and a good strategy for life. Because these molecules will spontaneously make an orderly barrier, creating an inside and an outside.

Oparin has shown that is all you need to have order; just the difference between in and out. He has taken just such simple spontaneous yet stable collections of molecules (he calls them coacervates) and given them a chance to impose order. They do, and for no reason other than they separate in from out. Oparin set up the same chemical reaction inside a coacervate and in the solution in which the coacervate was floating (the outside), he found order; inexplicable and unpredictable, but order. The rate of the reaction inside the coacervate was different from the rate of the identical reaction outside. Sometimes it would be faster, sometimes slower. That doesn't matter. As between men and women, it is only the difference that matters. *Vive la différence!* In both cases, it can lead to the creation of life. The coacervates are not living systems, so the "difference" is mystifying. All we can say is that the inside just has a certain *je ne sais quoi*, that's all. Not much, but it's all we need.

Once a system existed, and it really could have been as simple as a drop of oil, there could be order. Then the links in the chain from the sun could be forged. We think this chain was forged according to the Law of Horowitz, which states that any organism that can convert available compounds to a necessary compound can, not surprisingly, survive in the absence of the necessary compound. To explain how this has given us our totalitarian chemistry, let us return now to those thrilling days of yesteryear, when you were a cell and I was a cell in the Pre-Cambrian slime. Even further than that, to a time before we were cells. When we were what could be called *Ur*-cells—not quite cells yet, but primitive, incipient experimental precell models, just simple, molecular systems. There were different models of these *Ur*-cells, but they all required the same compound to survive. Since everyone was using it, there came a time when there was not enough to go around. That's crystal clear, but apparently you need some sort of magical spectacles to see it; it doesn't seem to make sense to a large segment of the population. If you use it up, it's not going to be there any more. Apparently a hard lesson to learn. To return to the slime. If one of these inchoate systems paddling around had been wily enough to carry around a clever protein, an enzyme that can take some substance available and convert it to the needed one, then what care that wily *Ur*-cell. If there is no more of the needed compound, it can whip it up on its own. The others, the ones that did not have the good luck or foresight to evolve such an enzyme, like the foolish maidens who didn't save their oil, would be in a very bad way. It could have been a fatal error. Is this a warning? Well, you know what happened to those unlucky *Ur*-cells.

This is how the first link in the chain may have been forged. What had to be there besides the enzymes was some kind of genetic system. Not much of one, really, just enough so that an *Ur*-cell could reproduce its biochemical success, make sure that there would always be more of these *Ur*-cells of superior type. Having a genetic system, like being born a Rothschild, would have been a fairly certain guarantee of worldly success and survival; survival reaching even unto many generations thereafter, because it gave life a pattern to follow and life thrives on predictability. Without a package of instructions on how to make the enzyme, the enzyme would disappear, perhaps forever. Too chancy. As time went on, another link was added to this chain, blueprints for its construction placed in the genetic system, and then another link could be added. Molecule A is needed, absolutely essential. Our first wily *Ur*-cell can convert B to A. When there is no more A, there are going to be no more *Ur*-cells, except for those a little less finicky about what they will eat. More time, and another enzyme; one that can convert C to B, which can of course be made into vital A. Now this *Ur*-cell has the hang of it, and adds another enzyme, one that converts D to C. If A, B, or C are all gone, that's all right, Jack; this particular *Ur*-cell can take care of itself. At this point, what a system would need to become an Horatio Alger hero was flexibility and reproducibility. The first made sure the success would be engendered over and over. And that, once it had been done a few hundred thousand times, is how we got our biochemistry.

This is actually a tale of two biochemistries; early on, biochemistry was responsible for an important innovation. If the substance you require is not available and you do not have the biochemistry to make it available, all you have to do is find something else that can make the conversion and eat it. This was the invention of food. Also, of predators and prey, of killing another to ensure survival. Quite an admirable trait to have picked up so long ago. We haven't let it go stale, freshening it up through the years with some interesting variations: killing for pleasure, killing for passion, killing merely because we are told to. The protists were the first with it, but we can, no doubt, proudly and justifiably claim to be best at it. Ardrey claimed we developed our desire to kill while we couldn't go for a walk without skinning

41

our knuckles. Lorenz claims we got it while we were still just a mammal. Really, we got it when it was decided that we weren't going to be plants.

Anything without a plant's biochemistry cannot make its own food; it is forced to become a killer, a parasite, or a necrophile; a delectable choice. We can kill other animals, or we can kill plants. We may have "got" ingratitude at this time along with our biochemistry. Do you realize what plants do for us? Everything, that's all. Not only do they make carbohydrates for us out of sunlight, they get rid of the poison, carbon dioxide (our biochemical toxic trash), and they fill the air with pure oxygen for us to breathe. They are also the one and only source of proteins, and give us almost all of the vitamins we need so desperately but cannot make for ourselves. Recently a famous young woman "of the acting profession" claimed she became a vegetarian so she could look animals in their eyes. She has it all wrong. I am a dedicated carnivore (a "meatarian" as a young friend has put it). What has that pig, cow, or chicken (especially the chicken) ever done for me, or for anyone compared to the gifts I have received and receive daily from my friends and suppliers, the green plants. How sharper than a serpent's tooth. To be a vegetarian is to be ungrateful.

We call the ones that can make their own food "autotrophs" (self-feeders). Autotrophs, because of their chemical talents, are automatically condemned to be food for others. The eaters are "heterotrophs" (other-eating). That is an ageless legacy we can't escape either. Our society—no doubt every society—is composed of heterotrophs and autotrophs. Always has been; some members nourished and surviving only by virtue of the talents of others, to the expense of the talented ones. Pandora had nothing to do with it. It was the protists,

42

those insidious inventive cells. They not only opened the box of human ills, they packed it themselves. To be fair, they also invented the benevolent heterotroph; not all heterotrophs devour the autotrophs that nourish them. Some become hosts in a symbiotic relationship.

Symbionts are permanent guests, providing something essential for their host's survival. For this, they are sheltered and sometimes nourished by their hosts. Symbionts are not parasites. As for their human counterparts, we use the term parasite to describe one who takes and never gives. Sometimes the parasite can take more than the host can give. The host dies. Symbiosis is far more pleasant and more interesting to me than is parasitism. I think of parasites as unlovely creatures, lacking in moderation: leeches, ticks, tapeworms, pinworms, fleas. Symbiosis can be beautiful. It can be morally beautiful, for symbionts achieve the nonexploitative relationship so sought by women of raised consciousness and by Marxists. It can be aesthetically beautiful as well.

My image of a symbiotic relationship is *Paramecium bursaria* and its internal symbiont, an alga *Chlorella*. Through the microscope it is a beautiful relationship: *Paramecium* golden, transparent and candescent, its body covered with thousands of hairlike cilia, beating in sensuous waves. Inside, it has hundreds of tiny glowing roses, emerald green shining with refracted light. (I always thought they looked like roses, other observers thought cabbages, and others single-celled algae.) The *Chlorella* are green plants; they can photosynthesize, transforming light energy into carbohydrates. *Paramecium* will eat them, but does not digest them; they are harbored safely inside. If deprived of their faithful, lifelong companions, *P. bursaria* will usually languish and die. This isn't merely a petulant decline, a pining for their devoted companions; *Paramecium* dies because the algae supply them with an essential substance. They will recover from their apparent brokenheartedness if they are given rich cultures of bacteria to eat. As some Roman poet-or-other once wrote:

> We can live without love
> What is love but repining?
> But where is the one
> Who can live without dining?

The lichens on trees and rocks have learned the same lesson that *P. bursaria* has: there are advantages to union; association can lead to mutual benefit. *Paramecium* must offer *Chlorella* something in return, because even free-living, normally independent species of *Chlorella* will happily take up residence inside a *Paramecium*. Lichens totally immerse themselves in their symbiotic relationship. A lichen is really two completely different organisms—two simple plants; an alga and a fungus. When they decide to cohabit, they undergo a metamorphosis in life-style and in appearance. It took biologists thousands of years to penetrate their masquerade. The two members in the lichenship can live apart quite well, and often do. But when times get rough (especially when it gets extremely dry), they get together, exchanging their independence and former identities for a single new identity and survival.

Many organisms have found happiness and survival in the nonexploitative haven of symbiosis. Termites and their symbionts are probably the *nonpareils* among such associations, the pale unattractive appearance of a termite houses positively Byzantine relationships. Termites live by eating wood. They eat it, but cannot digest it. For this, they keep populations of certain protists in their guts. Without them, termites could get no more nourishment from the wood they eat than we could get by gnawing on a pine tree. Only the protists can break down the cellulose walls of the wood cells; the waste products of this breakdown are what the termites live on. These protists are attractive as well as useful, decorated usually with long waving plumes, or tufts, of protist hair (flagella). One of these long-haired beauties is *Mixotricha paradoxa*, and it leads a double life. It is both symbiont and host. What seems to be *Mixotricha*'s own flowing mane, the envy of any Melisande or Rapunzel, is actually hundreds of other organisms, spirochetes. These are permanent parts of *Mixotricha* and provide it with locomotion. The Mixotricha have a few flagella of their own, but they use these just like rudders, for steering only. The rapidly beating spirochetes are the galley slaves who propel it through their constant efforts. This is not the end of the story; there is yet another convolution. At the place where each spirochete is attached to *Mixotricha* we find a small bump. On one side of the bump is the spirochete madly flailing away, and on the other side at the base of the bump is snuggled, permanently, a single bacterium. The last of the set of four Chinese boxes. Why it is there we don't know. They save energy by getting a free ride around a termite's gut, but they could just be tourists. If only our microscopes were powerful enough, we could see them, guidebooks to the termite gut in hand, cameras slung over their shoulders, shamed into overtipping by the already subsidized spirochete gondoliers.

THREE

Cellular Evolution

(courtesy of Prof. Karl Grell)

Plate V. *Allogromia laticollaris*. A single cell makes
these hexagonal plates of glass (silicon) and lays them
together with exquisite precision.

47

E ach of us is a nation of immigrants. Our cells, during their formative years, offered refuge and a new way of life to very talented citizens of divers and sundry origin. Like many of those immigrants 60,000 billion times larger, these enriched their adopted homeland, made important contributions, and became an integral part of it. Yet they stubbornly retained subtle but unmistakable links with their former identities and origins. Eukaryotic cells have many talents that the prokaryotes don't have. The best explanation for the vast dichotomy between them is that a number of independent single-celled organisms chose to camp out in a eukaryotic cell whose only talent originally was a nucleus. At first the relationships were just simple, straightforward, and symbiotic; both cells kept their independence. Gradually, however, the guests lost their own identities. Finally, they were no longer themselves but just one part of the complexity of their hosts, incorporated without being devoured. This was a point of no return—never could they regain their former freedom or selves. It was too late, the lives and destinies of host and symbiont were from now on inextricably entwined.

This is all a fairy tale according to many cell biologists. I am almost positive it is not. To show you why I believe in this one fairy tale even though my childhood, like the independence of the symbionts above, is gone forever, I must take you on an architectural tour, be a *cicerone* of the cell, but giving a too quick and inadequate tour. The cell in all its beauty reminds me of Ephesus, overwhelming, exquisite city. Ephesus is one of the few cities I truly love, dead or living. I love it, as I love the cell, not only for its clean, comprehensible beauty—the pale marble walls and pillars washed golden and rose by age and light with the blue Aegean of legend beyond the silted-in skeleton of the harbor—but also because of the thought that went into its construction those several thousand years ago. Ephesus, though dead, is a comfortable city, as memorable for the intelligence and practicality of its planners as for its overwhelming physical beauty. As the sailors of long ago left their boats in the harbor, they stepped immediately onto a long, gleaming marble avenue lined on both sides with memorial statues, many and impressive. It was designed to say to them, as they made their way up this great white way, blinding in the Aegean light, "Sailor, you are now in Ephesus, grand, proud, and beautiful." And, right at the end of this long promenade, the traveler would find spacious public baths and public latrines (with smooth, carefully carved marble seats). A wise, sanitary, and very humane gesture. (Are you listening, Los Angeles, Boston, New York?) Clean and refreshed a visitor had only to turn right as he left the baths and walk a few steps to satisfy other needs, deprived on a long sailing voyage. A large civic brothel was right there, and directly across the street was Ephesus' famed library. How functional. How thoughtful. A tour of our immigrant-receiving metropolis, the cell, would be as striking for both function and beauty.

A cell always leaves the same first impression. It is incredibly crowded in there; a welter of structures crammed together like rush-hour riders in Tokyo or New York subways, with no apparent breathing space. But each of the structures is separate, partitioned off by a membrane boundary. And each has an exotic name worthy of a Russian countess or a Balkan intriguer, except for the nucleus. Despite its comparatively prosaic name, like the unassuming nanny who turns out to be the head of the espionage network, the nucleus is the structural and actual center of the cell. This crucial compartment, where all of the cellular plans and plots are laid, is ringed by a double membrane, and it holds the genes, the ultimate dictators of the cell, wound into the coils of the chromosomes. And only when the cell must divide will we see a spindle of long thin fibers, attached to the chromosomes. These fibers will contract to pull a complete set of chromosomes into each new cell. At any time in the cell, we find a number of little sausage-shaped mazes, the mitochondria, holding not a miniature Minotaur, but a mind-boggling system for getting energy from sugar.

There are many more points of interest, but we are not going to stop at them. In, out, around, under, and over the rest of the structures of the cell is the endoplasmic reticulum: fold upon fold upon fold of membranes, a continuous communication and distribution system that reaches from the outside through the inside all the way to the nucleus. Only in the plants, though, are there the shining magical peridots, the chloroplasts that ensnare the sun in their green chlorophyll. These are the Merlins of the cell. They are wizards who can take the sunlight we see, feel, think about but cannot weigh, and change it into sugars. This, I am sure, is a far harder trick than turning a prince into a toad. Some princes have a very short way to go. Sugars, after the transformation, not only weigh something, but like sucrose can rot teeth and cause diabetic attacks in the bargain. This is only a temporary evil spell, however. The mitochondria can break the spell and change ugly sugar into beautiful, pure energy. Entropy-defeating, order-maintaining energy.

Now, according to the fairy tale I believe, the mitochondria, the chloroplasts, and a basic system of cell movement (part of which is the contracting spindle that distributes precisely the right number of gene-bearing chromosomes) all were once prokaryotes that started a nonexploitative relationship with a eukaryotic cell. But, as such things are wont to do, it went too far and ended up with one partner sacrificing its freedom. Why? Why does anyone do it? Security is the answer here as in the larger world. Security and a guaranteed income. There they are inside us, and because we are multicellular, we have more control over the environment; we guarantee them a constant pressure, constant temperature, protection from environmental vagaries of every kind. Most of us (at least those reading this) know where our next meal is coming from and thus give these squatters in our cells a constant, predictable supply of proteins, carbohydrates, and fats. A life of leisure and predictability far beyond the wildest protist dreams. The life of a single cell, alone, on its own, is perilous beyond all imagining. The only way to make sure the puddle in which one is swimming does not dry up and will always provide an adequate, well-balanced diet is to make sure that the puddle one settles in is a multicellular organism.

Most cell biologists, if they don't turn smartly on their heels and walk rudely away when this idea is explained, indicate, with varying degrees of politeness, that they put the idea at the same level of credibility as Cinderella or Jack and the Beanstalk. It isn't a new idea; it's rather old and has been suggested by eminent and respected biologists of the last century and of this one (including Nobel laureate Joshua Lederberg). In the last century, microscopists were struck by the integrity and occasional independence of some of the structures in the cell. They felt that these complicated physical systems could have been permanently added to the cell during evolution through hereditary endosymbiosis; which is actually nothing more than the simple idea that symbionts can in some cases be passed on from generation to generation, becoming as heritable as the family silver or Great Aunt Hyacinthe's moonstone lavaliere. Recently the idea has been revived, and supported with cold, hard facts gleaned from modern biology.

The nearly unanimous popular choice—the alternative explanation—is that cells evolved in a straight line, from prokaryotes to eukaryotes. Structures with complicated, specialized functions were merely enzyme systems (evolved according to the Law of Horowitz) that had become associated with membranes and had been coincidentally separated from the rest of the cell; a single cell, on its own, supposedly developed these many complex sets of biochemistry and special talents.

One trouble is, this theory postulates a large number of "missing links" that have never been found, and a complicated history of additions and losses of these cellular structures. The distribution of these talents (those of the mitochondria, chloroplasts, and contracting system) among the major groups of living organisms—plants, animals, protists, bacteria, and blue-green algae—simply makes no sense unless we view natural selection as forgetful, whimsical, and sloppy. It is none of these. It may leave an appendix hanging around with no apparent function other than to provide surgeons with fees when heart transplanting is slow, but in general it is a ruthlessly efficient force.

In the book that has started the All-New Hereditary Endosymbiosis Revival, Lynn Margulis's *The Origin of the Eukaryotic Cell*, we find that Goethe was right: *"Die erste liebe ist die beste."* The first theoretical love of cell biologists seems to be the best; the theory of inherited symbionts is completely consistent with facts (and the missing links are alive and well for everyone to see) and it makes exquisite logical sense. Margulis and a few others (a clear-thinking few to show how my bias is cut) have found arrayed against them a vast hostile host in closed ranks: most of the cell biologists. They are, henceforth, the other camp. The other camp is vociferous, powerful, and in one instance, quite vulgar in their denunciation of Margulis and the theory she has revived. You would think by their acrimony that she had proposed compulsory general practice of a combination of incest, bestiality, and matricide, rather than just a possible explanation of an important evolutionary event. An explanation that makes the most wonderfully simple sense, the way Darwin made sense. As shall be revealed to you, dear reader, the alternative explanation requires more willing suspension of disbelief and wilder leaps of illogic than any fairy tale my mother ever told me. Why do they refuse to listen to the sense Margulis has made? Don't ask me. Human beings are capable of an incredible number of vagaries and eccentricities, not to mention downright foolishness, that I can never understand, much less explain. I can't tell you why certain cell biologists are so refractory. And I don't know why people watch game shows and wrestling on television or prefer their cottage cheese with catsup, either.

Academic controversies are usually as boring to me as they are to those lucky enough to have escaped an academic career. But I think this particular one is more than the usual mad swirling of tea leaves in an academic pot. I think it reveals the truly average mind of the average scientist. The same scientists who are supposedly interested in finding out what is so. Among this group, if anywhere, we should expect new theories to get a fair hearing, for experimental science exists only to test ideas. Therefore, experimental scientists should be the segment of the population most receptive to ideas. Ha! And again ha! We are meeting with a paradox. The scientist, we are told, chooses what to believe in not by whim or personal prejudice, but by what is revealed through The Scientific Method, which reduces all personal bias to zero. But all scientists are human, most are intellectually conservative, and most are usually wrong (and very slow learners in the bargain). They were wrong about Copernicus's ideas. And Galileo's. And Pasteur's. And Mendel's. And Lister's. And Darwin's. And Goddard's. Yet the majority of scientists still haven't learned that the new ideas held by general consensus to be the most wrong usually turn out to be the most right. I except the nuclear physicists; they seem to do all right by new ideas—after all, believing in a quark with charm and color deserves a lot of respect. Biologists, though, on the average, are about thirty years behind their own field. I'm not asking that they love every new theory and clutch it warmly and wholeheartedly to their collective lab-coated bosom; I would be happy if they would just act like scientists. It is a bizarre paradox we are facing, for we find that experimental scientists (who are supposed to be fair) at times make the Spanish Inquisition a model of fair hearings and unbiased judgment.

There are some biologists who do feel that the idea of hereditary symbiosis is an interesting and convincing one, at least worthy of careful consideration before it is rejected. One of these is the "father of American ecology," Professor G. Evelyn Hutchinson of Yale. He was the first in the United States, far back in the thirties, to suggest that if we disrupt one factor in an interrelated environmental system, we can disrupt the whole system. It doesn't sound like a particularly crazy idea, but it took at least a quarter of a century before most people decided to agree with him. We were sorry (and still are) for that particular recalcitrance. Hutchinson, even though he isn't a geneticist (or probably precisely because he isn't) was one of a very few who realized that DNA had been identified as the genetic material in 1944. It took the geneticists until 1952 to arrive at the same conclusion (but then genetics is a very rapidly moving field). I'll stick with Hutchinson; his track record is better than that of biologists in general, and though he has clairvoyance, he hasn't a Ph.D.—a sure sign of trustworthiness.

I must admit that I have a personal prejudice that biases me in favor of Margulis' case. I happen to like logic; I think it works more often than not. The convolutions and involutions required to fit the alternate theory into the facts as we know them reminds me of one biologist's response to his student's completely illogical explanation of a phenomenon. "That's not right . . . (pause) That's not even *wrong*." It also seems to me perfectly peculiar that the other camp has such complete and revelatory certainty that cellular evolution did not involve hereditary endosymbiosis. They can't possibly know the cause they espouse is right. Admittedly, this shouldn't keep anyone from believing something, but in this particular case the events about which we are arguing happened billions of years ago. No eyewitnesses are still alive. We probably will never know. (I'm hedging because I have found too many times that nothing is impossible, we just don't know how to do it yet.)

Both camps propose that the prokaryotes were the starting point for cellular evolution. They are truly primitive and truly deprived; they don't have a nucleus, mitochondria, the contractile system or chloroplasts. The blue-green algae evolved from the prokaryotes; they photosynthesize, they can eat the sun, but that's about all they can do. They lack all of the other eukaryote status symbols. Some prokaryote invented the nucleus, organized its genes into the easily maneuvered chromosomes and gave them privacy for their highly technical work by barricading them behind a double membrane. Whoever did this was no longer a prokaryote; it was a protist.

This is when the trouble starts. Our camp feels that the prepossessing protists were parents to both plants and animals, for this group developed and tested everything the well-outfitted eukaryote needs, including the chloroplasts. The other camp is fairly sure the blue-green algae gave rise to green plants, in spite of the fact that they don't have anything (except photosynthesis) that plants have. They are willing to concede that animals came from protists.

For the other camp to be right—or even reasonable—identical versions of the nucleus, mitochondria, cell-division mechanisms, and contracting systems for locomotion, as well as the chloroplasts (carrying identical photosynthetic techniques) would have had to evolve separately in both plants and protists. Parallel evolution like this has happened. Birds have wings and bats have wings and butterflies have wings and pterodactyls had wings. Octopuses have eyes and human beings have eyes. The wings do the same thing and the eyes do the same thing, but there are obvious structural differences between these eyes and among these wings.

The eukaryotic skills are so incredibly complex that it is almost folly to imagine their evolving even once. To propose that they have evolved not only twice, but identically twice—well, that way lies madness. Aside from their structures, the mitochondria and chloroplasts each have a biochemistry of several dozen reactions. Identical long and intertwined chemical chains —nearly exact replicas—are found in any mitochondria or in any chloroplast with chlorophyll, from plant, protists, or animal. A belief in hereditary endosymbiosis doesn't require much believing. Just that each of these extraordinary, improbable associations of structure and biochemistry evolved but once. They were then acquired by the eukaryotic cell in total, as symbionts who had developed these systems took up permanent residence, tired of trying to make it on their own, in a protist of some sort. This is an infinitely more *possible* explanation, and far more simple than any other so far proposed. You will have to take my word for it (and the word of Bucky Fuller), life evolving has held one tenet close: "Simplicity is the essence of design." It is far too expensive and wasteful to be needlessly complicated. To be wasteful in the face of such fierce competition as life offers is to be stupid and to be dead. Simplicity is survival.

What perhaps convinces me most that cell evolution liked the idea of hereditary symbiosis even if many biologists don't is the cold, hard fact that each of the three systems—photosynthesis, the contracting fibers, and the mitochondria—has its own genes. Their *own* genes, not related to the genes of the cell where they are found. These genes—especially those of the mitochondria and the chloroplast—are there to run the individual structures. This is not conjecture. All three have been seen to behave independently, in some instances completely disregarding what their cell was doing. They have wills of their own. If they had evolved gradually, as just systems of enzymes that accumulated and then were wrapped in membranes by the cell, why would they need their own genes? Where would they get their own genes?

CELLULAR EVOLUTION

Unless you belong to the other camp, the answer is obvious. They have their own genes because they once had to have their own genes because they were once completely on their own. I like facts and I like logic, but I also like the story of the cell according to Margulis because I am delighted and intrigued by the idea that I am a New York City in my own right (cleaner and a bit more solvent, though); the Big Apple on a small scale, hosting and sheltering my own array of immigrants. Mine come from only a few different ethnic backgrounds, but I have no doubt they contribute immeasurably to my existence, providing me with a variety of experience and ideas. I realize I don't have any alternative; they were there before I was. But, having immigrants for grandparents, for whom America was, always and ever, protective and generous (for them the "Liberty Statue" kept her promises), I feel that the least I can do is to be as gracious and welcoming a homeland as America was to Constantis Rickis and Anna Luksyte my emigrating grandparents.

Our personal immigrant epic began when the first immigrant-symbiont arrived. Our own prokaryote Prometheus, bringing with it the invaluable talent of being able to use oxygen to burn our food up more efficiently. Oxygen is a universal poison, toxic to all living cells. It simply isn't our element; we evolved in its absence. When some cells discovered photosynthesis, their waste product from feeding on the sun was oxygen. Their filling the atmosphere with this poison gas was a great shock to the rest of the living world. As in any crisis, this emergency meant different things to different beings (all only cells at this time). Those who could not cope or figure out how to detoxify the poison simply died; others went underground into the airless mud or slime of swamps, where we can find them still. For those who had flexibility and ingenuity, this was their big chance. One way of dealing with this deadly peril was to light up; the luminescence of certain protists (*Noctiluca*), fireflies, bacteria, and mushrooms began as a detoxification program. Free oxygen in the cell will form hydrogen peroxide, useful if you want to live your life as a blonde and nature has decreed otherwise, but quite horrible if you merely want to live your life as a cell. It is a fierce poison. By lighting up, however, peroxide is converted to harmless, utilitarian water. The little beetles we call fireflies recognized the aphrodisiac value of the chemical byproduct of this reaction, light, and shading it red, orange, yellow, green, blue or white, use it to signal their intentions at mating time: "Male *Lampyridae* of certain species seeking female of same."

Other cells figured out how to use the oxygen to become more efficient at getting energy from sugars. Before this, everyone had to ferment to get the energy that photosynthesis puts between the carbon atoms it joins together to make carbohydrates. The energy can only be retrieved when the bonds that bind the carbon atoms are broken, and fermentation can only go so far in the breakdown; sugar (glucose, actually) is stripped of its carbon atoms, one by one, in one of the most pleasant and nourishing chemistries ever evolved:

(fermentation enzymes)
glucose ———→ alcohol and carbon dioxide.

This lovely piece of biochemistry provides our jugs of wine and loaves of bread. (Less romantically, gin, whiskey, and vodka as well.)

Yeasts are terrific fermenters. They use the sugar we provide them in bread dough to carry out fermentation; the alcohol bakes away and the carbon dioxide makes the bread rise. Other yeasts (hundreds of species) form the modest gray dust on the grape (the "must"), the particular species is determined by the variety of grape and the geography. The grapes are crushed, the sugar released, and the yeasts, bless them, go to work. The yeasts are "Bacchus, ever fair and young, who drinking joys did first ordain." It is infinitely satisfying to cell fanciers that Bacchus was a fungus, a unicellular fungus, and not a puerile, rowdy youth with vine leaves in his hair. Yeasts ferment the sugar to alcohol. They stop when they have converted the grape juice to about 10 to 12 percent alcohol. Under these conditions they can work no more. If all of the sugar is gone by this point, the wine is dry, and these infinitely clever fungi give us Burgundies or Bordeaux. If there is still sugar left, we have Tokays and Muscatel. When the fermenting wine is kept inside a bottle, the carbon dioxide provides us with Champagne. What delights and surprises the prokaryotes were preparing in the pre-Cambrian slime.

Fermentation is truly a kind, beneficent legacy from those ancients, but delicious as it is, it is not efficient. It is actually a sloppy, wasteful way of getting energy. Only about 7 percent of the total energy available in a glucose molecule is released for us through fermentation. With the help of oxygen (and many dozen additional chemical transformations), glucose can be burned up more completely, and we can get more than half the energy out. Any cell that could not only tolerate oxygen, but could learn to use it, was automatically wealthy, like someone with an infallible set of hunches about the stock market—investing the same amount as other speculators, but getting vastly greater return. And these cells spent: on growth, on building and maintaining fancy structures such as the membrane around the nucleus, mechanisms for movement or chromosome distribution, for transmitting nerve impulses, and all of the other qualities that make our cellular countries great. It was money well spent. Blue green algae lack these structures and the talents that come with them. They can't afford them, even though they have photosynthesis, for they do not have the sugar-smashing mitochondria. Neither do the prokaryotes. And we all know where the blue-green algae and prokaryotes ended up. If we believe that multicellular green plants came directly from the blue green algae, we have to believe that mitochondria, exact in every detail, arose twice, once in the plants and once in the protists. I ask you, do you find that easier to believe than the proposal that mitochondrial capability evolved only once, in one kind of cell, and then once and only once took up residence in a protist?

We know that symbiosis occurs all the time and that the protists are a specially hospitable lot (from the western United States or from the Middle East, no doubt); we know of at least a dozen protists who have turned into residence hotels for algae, bacteria, and even for other protists. We're not asking too much in asking you to believe in hereditary endosymbiosis; we even know of symbiotic relationships that result in the ability to tolerate and use oxygen. We think that just like these, a cell with this oxygen talent moved into a protist, and after millions of years, became what we call a mitochondria.

Then came another stranger with another contribution, a set of protein fibers that could contract, that could push, pull, or propel with a whiplike motion. No longer would the protist have to content itself with sitting like a bump on a log, or with the characterless ooze of the amoeba. It could move! Really move. Just about every animal and higher plant uses exactly this system for propulsion during at least one stage of their lives. It's the way human sperm gets around; whipping a long tail of these fibers back and forth, it can zip along at about five centimeters a minute. Nothing to set Indianapolis records, of course, but it serves. It obviously serves. Even plants have mobile swimming sperm; it is quite exciting to watch fern sperm (with three whiplike propellers) thrashing madly about in all directions in a desperate attempt to live on into the next generation. Even ferns have ambition. And all higher plants and animals use the same snapping fibers to separate their chromosomes at cell division. Every one of these fibers—whether propelling a fern sperm, a human sperm, or moving *Paramecium* rapidly around—have the same characteristic structure; an unmistakable pattern. Each of them is composed of a complicated spiraling rosette of nine doublets or triplets of fibers. The structures that direct the motion of chromosomes in all eukaryotes, plants, or animals, show the spiraling nine. The other camp asks me to believe that all of these identical structures evolved at different times in different places. Fat chance, I say.

It doesn't make sense. What does make sense is that some cell had these fibers, maybe a prokaryote like the spirochete galley slaves of *Mixotricha,* and just like this inseparable pair, the symbiont provided motion for its host. And even more. It allowed its host to use the same fibers to make sure that when it divided, an identical set of genes was given to each new cell. A valuable skill indeed. The amoeba *Chaos chaos* does not have a foolproof way of sending genes into its offspring, and it has had to compensate. It has hundreds of nuclei, hundreds of sets of the genetic encyclopedias;

when it divides, at least one full set is bound to be passed on. And the immigration continued yet a little further. One of these protists now equipped with contracting fibers and mitochondria opened its arms (or mouth) to an algae, just as *Paramecium* gave a home to the beautiful *Chlorella*. These sun-eating guests became the chloroplasts.

The hospitality of the protists is phenomenal and unending. They have been kinder than almost any country has been to its immigrants. Their hospitality brought them success. With all of these acquired talents, they were able to turn into multicellular animals and plants, able to construct the complexities of a flower or a nervous system. They had the energy for motion, the structure for the giant redwoods. Understanding that it was eukaryotic talents that led to multicellularity, we can solve one of the great evolutionary puzzles: the heretofore mysterious but complete division between the Cambrian and Pre-Cambrian eras. In the Cambrian (before about 4.5 billion years ago) there were very few organisms that could leave fossils. Then all of a sudden (a geological "sudden" is about 40 million years), there was a great explosion of life, at least of life that could leave their stony shadows in the fossil record. A surfeit of fossils and a surfeit of forms. Almost every major group of animals can be traced back through the family tree all the way to this explosion at the opening of the Cambrian era. The other camp can't explain this Cambrian blossoming and blooming of life. Margulis can. The Cambrian was the "Age of the Cell," when all of the great leaps forward we have been talking about took place; when these had all been accumulated—mitochondria, contracting fibers, chloroplasts, nucleus—then and only then could the "Age of the Organism" begin. Only at the end of the Pre-Cambrian did single-cells have all of the skill needed to make bones, shells, muscles, eyes, tentacles, leaves. To put together octopus, fish, sharks, ferns, and pines.

Finally, we must consider the result of these migrations into our cells in personal terms. Plants carry four separate sets of genes, and we carry three. Three sets of independent operational instructions. We certainly can't live without our mitochondria, nor reproduce without chromosome-separating fibers and the whipping tail of the sperm. Now, who's really in charge here? Are we the willful, precipitate creatures we think we are, or merely a well-put-together soup kitchen, providing free food and shelter for these hangers-on? Do we have 60,000 billion cells just because these little creatures within us needed the stability and mobility that size and complexity could give them? Are we merely a high-rise apartment house, the built but not the builders? Sleep well.

In any event, it was wise of these guests (or rulers) not to call attention to themselves. *Homo sapiens sapiens* does not have a record that inspires confidence when we look at how the species treats anyone smaller or weaker than itself. We can feel invaded or threatened at the drop of a hat. Perhaps, as Dostoevsky's Grand Inquisitor thought, there are some things that should be kept from humanity, for its own sake, to keep them from harming themselves or others, to keep them from despair. Who can imagine (knowing us), to what heights of witlessness, foolishness, and silliness we might rise, if we knew the true relationship between ourselves and our mitochondria? Everything from paranoia (to which we seem to be especially prone) to suicides out of spite by those who refuse to be a welfare state to nameless billions. It would never end. There would be worshipful cults, adoring the invisible deities that guide our destinies; seances and trances to try and establish a dialogue or some kind of communication with these beings that exist in another plane of existence. New pseudo-sciences such as mitochondrial forecasting, for are they not closer to us and in more real contact than the stars upon whose courses so many assign their fate? We are desperately searching for the answer to why our lives run as they do. We are pretty sure there is not someone with a long white beard there anymore, even if he was there once. To fill that void we try gurus, stars, the palms of our hands. Let's face it, our mitochondria and contracting fibers are as vital as any of these and are equally as good candidates for such adolescent adulation. I'm sure I couldn't bear it. Forget everything I have been talking about.

FOUR

•━━▶━━◀━━◆━━▶━━•

Locomotion

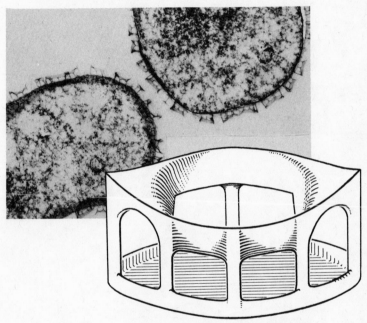

(courtesy of Prof. Karl Grell)

Plate VI. *Paramoeba eilhardi*, showing the tiny flexible
structures, used as suction cups, covering the surface.
"It is the unimaginable sound I love; the one that must
be made as these hundreds of tiny gumshoes are
squashed down, one after the other, and are pulled
loose again."

Ball bearings use something like 30 percent of all the energy used in the United States. With that phenomenal amount of energy going into just motion, I don't understand why satellite cameras don't show merely the blurred coastlines of a perpetually shimmying nation. Obviously we refused to be satisfied with the motion available to us through our cells. It wasn't fast enough and it wasn't easy enough. Our dissatisfaction led to the idea of fuels, of using someone else's energy rather than our own. We are now so unconscious of the limitations of the motion innately our own, that only to a rare person does the word "go" mean to walk or run. It means rockets, trucks, automobiles or ships. We gained some magic from our piracy of the energy other cells have made. We have explored every part of our earth, evaporated the unknown of our planet; there is no reach of it anywhere in any direction that has not at least been contacted. We have shattered some myths and learned our earthly mastery. This could have been done on our own energy. But once we knew our planet, fuels allowed us to circumvent our inability to tolerate boredom and to satisfy our driving, innate curiosity. We have been able to touch the cold, formerly unapproachable goddess of the moon. We have been able to send mechanical creatures when we could not go ourselves, to scrabble in the soil of a neighboring yet distant planet, seeking some even more distant biological relative. These are achievements unthinkable, unimaginable if we had remained satisfied with merely relaxing and contracting our own muscles or the muscles of some other beast.

We have lost some magic, too; we all know that. Anything but just muscular motion is too fast: too fast to see, too fast to stop and see. That is where our experiments with locomotion have brought us. We have won a good deal, but we have lost too; humanity's perpetual Midas touch. The myth of Midas is perhaps the story of mankind; more the study of man than the study of a man. The blessing of being able to turn all into gold at a touch is soon seen for the curse it is; when food becomes glittering and valuable but incapable of nourishing, and rest is sought but not found on beds of hard and unyielding fortune. Such is the burden of our sapience, the curse on the gold of our invention.

As always, the protists were the great inventors and experimenters in the field. They never came up with a wheel, however. The closest they could get was a mechanism that can spin a head on a protist body, or the body on the head—forty times a minute. This spectacular trick is managed by certain protists sharing the termite's gut with relatives that have their own unique form of locomotion, the ones that propel themselves by their symbiotic spirochetes. There must be something quite special about a termite's gut; it seems to have been one of the most productive think tanks of all time. This head turning is quite spectacular; the little beasts can do it for hours on end, and apparently to permit this continual twisting, the cell membrane actually becomes liquid. A completely astounding trick. This motion is very like a wheel, except that it can't continue indefinitely. Perhaps because they couldn't quite manage a continually rolling wheel (although I don't quite understand why not—it seems infinitely easier than triple concentric geodesic domes), we are on the moon and Mars, and they are only everywhere else. Yet their experimenting did lead to precisely controlled locomotion, and a system of locomotion, based on contraction of fibers, that has been almost universally adopted, fiber by fiber, in higher animals and plants. We use this system, but to reproduce, not to move. For movement we have adopted the principle (contracting fibers) but not the structure of this same system.

Yet this particular way of moving is not the only way protists move; there was a lot of other experimentation. Today's protists show us that these experiments were complex, interesting, and even incredibly charming; they led to failure nonetheless. Motion, however, is worth any amount of effort, any number of failures. To move: to hunt, to explore, to change environments, to get away from unpleasantness and to find comfort. The discovery of motion was a gigantic step toward being able to take advantage of the environment, rather than the other way around. Movement also brought the first faint glimmerings of learning, and of interpreting messages from the environment. Motion had to precede interpretation and understanding.

The motions of protists are sheer delight, yet their most frustrating aspect. It is impossible to verbalize an amoebic ooze, the ebullient bounce of a hungry *Didinium*; analogies are scarce and always strained. If, for this, we could just leap back into the eighteenth century, where almost every self-respecting house had a microscope, where the family would linger long and lovingly, fascinated by the world beyond their vision; when the attachment to one's microscope was intense and even passionate. The audience understood when the heroine of one drama refused her suitor's desperate pleadings with a fervent and indignant "What! And leave my microscope!"

It is sad that the day of such amateurs (that word does mean lover, you know) is past. Sad, but not particularly strange. The protists demand patience, and we are an evolving species. We are no longer *Homo sapiens sapiens,* wise and wishing to be wise. We are now *Homo sapiens instanter,* the "at once" species: instant food, instant knowledge, instant oblivion, instant transport, instant relief from annoying elements (divorces and nursing homes to instantly be rid of husbands, wives, or parents who refuse to gratify us when we want and how we want). We are now devoted to the gods of "as soon as possible." There is little we will not let our senses suffer in order to worship these gods; no toxic, tasteless trash with which we won't assault our taste buds, no noise too excruciating for our ears when we want to forget or to move fast. So the experience of protist motion is almost impossible for *H. sapiens instanter* to obtain.

But the rewards for patience, time, and practice might tempt even the most ardent devotee of instant. A drop of water from a warm and long-standing pond is so completely entertaining that it seems a masque or revel designed solely for our amusement, to entertain and astound us. There is no tinge of embarrassment, as there might be at watching protists mating. That can get quite personal. There is no voyeurism in watching a drop of pond water; we feel like an expected audience for feats of skill, acrobatics, and comedy.

The well-chosen drop of pond water is a circus, slightly surreal. Some of the movements are familiar; most are not. The first creature usually seen is an alga; simple plants stretching greenly in long, slender ribbons across the field of view. It is just a plant—a very primitive one—but watch. It moves. Swaying from side to side, gliding with excruciating slowness and grace across the circus of light. It is eerie *Oscillatoria,* and even a little frightening. Plants don't move, aren't supposed to move. Yet *Oscillatoria* does, sinuously, like a

torpid delicate snake, wind its way. No one knows
how. The slender green phantom glides inscrutably on.

The amoebas are slow as well, and subtly hypnotic,
but more comprehensible. There they'll always be,
seeping across the bottom, a perpetually leaky puddle
of protoplasm; but one that mops itself up as it moves
along. The amoeba truly is like a blob of jelly—not a
shapeless blob, but a protean blob of infinite shapes.
It changes its shape to move, and can move because
it knows the spectacular trick of changing its protoplasm from solid to liquid and back again at a moment's notice. Mostly, an amoeba is soft, fluid jelly, but
the external limits—the border of the cell—has been
converted into a more rigid gel; it is a skin of a pudding that can be constantly reset, changed immediately
back and forth, from pudding to skin, at any time.
Amoeba move by walking on their false feet (pseudopods), transient, flexible tubular extensions of their
own constantly flowing substance. An amoeba can have
as many or as few false feet as it likes, one region of
the solid cell limit liquefies, and liquefied amoeba
pours out, simultaneously solidifying to make the walls
of a tube through which more amoeba is flowing.
Amoebas, are the only creatures that can "ride madly
off in all directions," like Stephen Leacock's hero. They
can make and flow into several pseudopods at once;
true Geminis, they can be moving in several directions
at once. An amoeba never is torn apart through inde-
cision, though, for even if two parts of the amoeba are
inclined to go in different directions, a choice is always
made. We could interpret this as schizophrenia or just
confusion, but it could also be a judicious simultaneous
sampling of conditions, in order to make a wise choice
of future direction.

Most of the time the amoeba is watched from above,

and from that point of view its motion seems to be just a characterless ooze. The amoeba is more three-dimensional than popularly thought, however. We have only to look at one from another angle, from the side, and we see the false feet for what they really are: legs. Awkward and a bit elephantine, perhaps, oozing and rubbery, but still legs that reach forward and take steps. One false foot stretches out and over, then steps down to the surface beneath. The amoeba partly flows into this; then another foot is formed, lengthens, and takes another step. It is clumsy; a headless dinosaur lumbering along in awkward gymnastics, performing a multilegged cartwheel in slow motion. It moves like a tank in more ways than one. The interaction of the liquid moving over the solid part seems to be exactly like the wheels of a tank moving over its flexible metal track, extending short little arms, like a ratchet wheel into the solid part, to move along. The amoeba must be given the dignity it deserves: no matter how it achieves its motion, it does stand up and walk; no mean feat for a spineless, boneless creature.

One of the most memorable photographs I have ever seen is one in a book about bats by Alvin Novick. On one page there are several shots of a vampire bat, en route to its victim. Not wildly flapping and veering, as Bela Lugosi would have us believe, but very stealthily tiptoeing up behind. I am endlessly fascinated by this image of that fanged and maligned creature, stalking forward, its wings raised and partially folded, a dainty dowager fastidiously raising her skirts as she walks along a puddled street. This tiptoeing bat is exactly what a moving amoeba looks like, plodding slowly and silently towards its goal. Perhaps this is the way an amoeba also approaches a victim, substituting stealth for the speed it can never have.

LOCOMOTION

An ordinary everyday amoeba even when walking is no match for *Paramoeba eilhardi,* which has added a spectacular feature to simple amoeboid motion. The surface of this, my favorite among all amoebas, is covered with tiny, fenced, transparent eight-sided structures, flexible and open at the top. These are apparently used as suction cups, to get a firmer hold on the surface as *Paramoeba* steps along. It is the unimaginable sound I love, the one that must be made as these hundreds of tiny gumshoes are squashed down, one after the other, and are pulled loose. I am absolutely certain it makes the same noise as galoshes; the galoshes of an entire class of schoolchildren walking along a linoleum floor. It has to be completely unmatchable, hundreds of tiny sucking squishes, as *Paramoeba* wetly saunters on.

I also greatly admire the public rapid transit system developed by one protist, a species that tend to group together, rather than live separately on their own. By name, *Labyrinthula coenocystis,* this protist does build a labyrinth: delicate, beautiful, and practical. A colony of *Labyrinthula* builds a network of glistening transparent threads, strung in every direction. When a *Labyrinthula* wants to travel, it among all the protists has a choice of public or private transportation. It can just swim around on its own, or it can take the public system, moving swiftly along inside of the labyrinthine threads, like individual cars speeding through a vacuum tube. As it is in most of the exotic locomotions of the protists, the mechanisms of this eludes us. We are far too large to find out how *Labyrinthula* manages this efficient crystalline monorail, scooting and shooting back and forth through the threads.

Most protists have adopted a single system of motion. We find innumerable variations, of course, per-

sonalization of the basic system by the ever-individual-
istic protists. It is a flexible, versatile, efficient system
that, ever since the protists, has been adopted continually by plants and animals. This is the spiral of nine
sets of smoothly sliding fibers. The protists use it either
as a long or short engine of locomotion, putting it inside cellular extensions. The structure is identical in
both, but when long and whiplike it is called a flagellum (pl. flagella); when short it is a cilium (pl. cilia,
Latin for eyelash). Protists can have from one to a few
dozen flagella, and they can use them in a variety of
ways. The green, spindle-shaped *Euglena* carries two
flagella at the front, one long and one short. The longer
one is held to the side, trailing backward and slightly
coiled. To move, the *Euglena* simply sets up pulses of
motion through this flagellum that run from the base
to the tip, the same way we could flick a lightly coiled
rope or bullwhip. The flicking of this flagellum causes
a double rotation of *Euglena:* it whirls smoothly and
madly around on its own axis, almost too rapidly to be
seen. It also gyrates in larger circles, the front describ-
ing large circles as it moves through the water, rotating
around the tail end that stays fixed in a straight line.
Through its flagellum, *Euglena* is turned into its own
propeller blade, which moves it through the water just
as an outboard motor would.

Another protist shaped very much like *Euglena,*
equipped also with two flagella, moves in an entirely
different fashion. In fact, *Peranema* does not seem to
use its flagella for propulsion at all. One is wrapped
around the body in a spiral, attached at the front and
the rear; the other is held rigidly straight ahead, except
for the tip, like a ringmaster's whip. For this creature,
though, none of the gyrations of *Euglena;* it moves
smoothly and swiftly behind its unicorn horn. Its

76

smooth glide probably depends on the trail of mucus it lays down, and upon which it slides or skates. Another protist, *Gonyaulax polyhedra,* uses its two flagella to present an immensely charming picture. *Gonyaulax* is faceted and red, as attractive as the pomegranate seeds that tempted Persephone. Round its middle is a deep groove—a depressed girdle—and lying in there, beating, is one flagellum. Running from this girdle to the base of the cell is another groove, and in this, too, a flagellum sits and flickers. This helps make *Gonyaulax* a one-celled spectacular; it spins madly around like a top, moving forward as it spins. It also flashes on and off, lighting up as *Noctiluca* does. A frantically pirouetting, flashing ruby gem.

The use of cilia, the shorter versions of the flagella, is perhaps even more imaginative. Some protists have many, wearing a hairy coat of vibrating fur; others may have just a few dozen cilia perched awkwardly at the top of the cell, like a bizarre toupee actively self-grooming. The beat of the cilia can be used to do two things: to push the protist through the water, or to push water through or past the protist, if the cell is stationary. *Paramecium* is probably the most graceful of the protists with cilia, a slightly asymmetrical submarine wearing its silky coat of thousands of evenly spaced cilia. The cilia of *Paramecium* are synchronized into an overall asynchrony, so that the hairy surface of a *Paramecium* moves in waves, like the wind-blown swells in a sea of grass. The asynchronous waving propels them rapidly forward, and also turns them on their own axis. *Paramecium* are among the best-behaving (i.e., the most interesting) protists. They have recognizable behavior, repeated in every *Paramecium;* but since it involves moving toward or away from something, we know that motion had to be there before the behavior. A pleasant spot, to a *Paramecium,* is one, with a warm bath (80°) and plenty to eat (bacteria). They obviously cannot think, but they can perceive, and they can reverse direction. This is all they need to make themselves comfortable. It is simple, unthinking, stereotyped negative behavior, but it puts a *Paramecium* precisely where it would choose to be, if it did have the capacity for choice. Their waving cilia bring samples back to the cell of what things are like up ahead. They can tell whether it is colder, warmer, more acid (a sign of bacterial life). If it is colder or less acid than where the *Paramecium* is, the cilia reverse, automatically. The *Paramecium* backs up, and turns 30°, away from the side on which its mouth is. Then it goes forward again. If things still look bad (or just slightly worse), the cilia reverse again, there is another 30° change of course in the same direction as the first, then back into forward gear. Again, and again; if necessary, it will circle 360° if nothing ahead appears promising. This behavior, brainless as it is, has an enviable result we don't always manage by our highly cerebral technology. The *Paramecium* always ends up where things are better, simply because the cilia refuse to propel it where things are worse.

Amoebas, strangely enough, seem to be more perceptive than *Paramecium.* Amoebas, too, have motion as the only dimension in which to express behavior. They will pull their false feet back from where something unpleasant has been detected and push another foot out in another direction. If they find themselves surrounded by naught but unpleasantness, they give up, withdrawing themselves into calm, stoic lumps, refusing to extend themselves by so much as a single small pseudopod. If the bad conditions continue, they will philosophically adapt, extending after a while a

78

searching, seeking pseudopod, deciding to make do, to ignore the unfortunate vagaries of their existence. *Paramecium* never become this philosophical; they will frantically circle, trying to escape, no matter how relentless and inescapable their unsatisfactory environment seems. Amoebas also manage active pursuit of something that catches their interest, behavior completely alien to the *Paramecium*. To an amoeba, anything that is large and moves a lot is more interesting than something that is small and doesn't move much. An amoeba can obviously detect degrees of size and activity, for it expresses interest by the size of the pseudopod extended; the larger and fatter the pseudopod, the greater the interest. Only when there is something relatively large and actively thrashing about will an amoeba extend a large, high-interest pseudopod. Even with this exceptional amount of perception (or, really because of it) the life of an amoeba is doomed to be one of perpetual frustration. An amoeba must be the Tantalus of the protists because it is able to move only very slowly. The faster the prey, the more interested the amoeba will be, but the less able to succeed in its quest.

Blue *Stentor*, cerulean single-celled trumpet, can be counted among the protistan "very rich." If *Stentor* doesn't have everything that we might want, it still has more than any other protist has. It has beauty; although it is large, it is visible only as a spot of dull blue without a microscope, not visible in all of its considerable beauty. It is a single-celled replica of the morning glory, a flared trumpeting shape striped with alternating blue and white. Most of the time *Stentor* is immobile, stuck by its stalk to a surface, feeding through rows of rotating cilia that whirl currents of water into its mouth, sweeping in particles digestible and indigestible. If they are indigestible, the particles are rejected by a mere flick of the cilia, a reversal of beat; the same technique used by the highly mobile *Paramecium* to remove themselves from something undesirable. *Stentor* uses this common negative reflex to kick food out of its mouth. *Stentor* has been nominated by one protozoologist as "one of the most beautiful animals in existence." And that may be; it is truly lovely, this fringed azure trumpet. And it is, beyond a doubt, the most expressive, the most Gallic of animals, especially at registering annoyance. (While it is never apparent that a protist is happy, some of them can indicate beyond a shadow of a doubt that they are not.)

Like other protists, *Stentor* will seek out conditions they consider pleasant; they prefer the shadows, so one can actually herd *Stentor* with light, corraling them into a single point of darkness. But it is their behavior when annoyed, their capacity for communicating their annoyance and displeasure that gives them a good deal of their charm. If *Stentor* is purposefully annoyed, by touching it with a needle, or tapping sharply the glass slide where it is attached, it will first bend over to see whether the intruder tastes good. For a creature without jaws or teeth, *Stentor* can manage a fairly fierce bite; it can keep hold of a wildly struggling multicellular animal, or take a bite out of another *Stentor*. If the intruder seems to be promising cuisine, *Stentor* will try to suck it in for sampling. However, if the nuisance is extreme, continued tapping or a very vigorous poke with a blunt needle, *Stentor* turns away, just as most of us do if we detect an over-the-shoulder newspaper scanner. If the unpleasantness continues, *Stentor* will do one of two things. It can hunch up, pulling its mouth down into its shoulder in a disconcertingly human gesture. After a while, it may decide to brazen it out,

ignoring the nuisance and continuing to go about its
Stentor business. We could say it becomes accustomed
to the nuisance level, and "learns" that poking or tapping is not harmful. On the other hand, it can decide
not to be so tolerant and to hoist its anchor, using its
cilia to sail away. The course of action is determined by
security; if the food supply is rich, *Stentor* has to be
annoyed unmercifully before it will decide to leave.

Another lovely—but incredibly deadly—protist is the
trypanosome. Trypanosomes are shaped like eels, with a
rounded "head" and a long, tapering tail ending in a
flagellum. From head to tail waves a single fragile membrane, delicately undulating like a tasteful flounce of
chiffon. The flagellum actually runs the length of the
trypanosome, through the loose top of the membrane
like a flexible curtain rod. Fluttering its membrane se-
ductively, this fragile, delicate creature wends its way
sinuously through our blood, injected originally by
the tsetse fly. For the lovely *Trypanosoma* cause fatal
sleeping sickness; their waste products are to us deadly
poisons. Once inside our bloodstreams, they take up
residence in our nervous system, and a slow poisoning
begins. First the characteristic lethargy sets in, then
death makes a final unwakening sleep. The movement
of trypanosomes is sheer poetry, though extremely difficult to appreciate.

The motion of one other protist, though less poetic, is
far more innocent. These are the protists that fuse their
cilia into little paddles. The original admirer of the
protists, Leeuwenhoek, called these structures "paws,"
and perhaps he was right. For these creatures don't
use their fused cilia as paddles, they do indeed use them
as paws, to pad around on; they walk on their stiff,
sturdy little paws. They really do walk, looking the
most purposeful of all the single-celled creatures, with

a real "get there" air about them. They lack only an umbrella and a bowler to be tiny mimics of a bustling British clerk, as they march ahead with their hurried, no-nonsense, earnest gait. They will also vary their direction and their speed, turning smartly on their stilt-like legs if the present direction seems unrewarding, or if there seems to be something of interest to investigate.

It is hard to find an unattractive protist, especially any one in motion, although not all can match the Winnie-the-Pooh appeal of the *Euplotes* and its tiny paws. And as lovely as *Stentor* is, with its obvious blue charm and expressive Gallic shoulder shrugs, it still appears rather thick and ordinary compared to the delicately winsome *Vorticella*. *Vorticella* is pure *art nouveau;* glistening and rainbowed with refracted light, it is an exact living miniature of a Tiffany goblet I once saw. Both were capable of taking my breath away. They are both tulip-shaped and iridescent with the bowl of the goblet balanced on an improbably long, fragile stem. It is delicacy and balance carried as far as they can possibly go. This alone would be enough, beauty is its own reward. *Vorticella,* though, has an alter ego, almost as appealing but less sophisticated. There is a long thin fiber running through its transparent stem. This can contract and, when it does, the stem is pulled into a coil, and there *Vorticella* bobs back and forth on its glass spring, an exquisite and slightly comic jack-in-the box.

Cilia and flagella argue strongly for cellular evolution through hereditary endosymbiosis: progress through the permanent adoption of useful symbionts who have learned some advanced chemistry. Cilia and flagella are nearly universal. In every higher plant and animal except for a few aberrant rebels, cilia or flagella appear at least once during a lifetime. Both cilia and flagella are built alike, with the spiraling nine sets of fibers and this is a truly wonderful invention. The fibers are actually tubes, that lie parallel to each other and slide back and forth along each other. The entire structure can bend or beat, as the fibers lengthen and shorten by being pulled past each other. The same mechanism separates chromosomes when a cell divides, and just as efficiently as it feeds and moves protists. Since the time when protists were kings of the world, there apparently has been no better way devised to move a cell through a liquid, or a liquid past a cell.

We use both cilia and flagella ourselves. Males use the flagella. The all important tail of the spermatozoan is exactly a flagellum, a protist-developed flagellum, indistinguishable from the ones that move *Euglena* or *Gonyaulax*. Females use the shorter cilia. The same cilia that feeds *Stentor* moves the human egg to the uterus; the Fallopian tubes are lined with thousands of beating cilia. I wonder if the irony there means something. We need the motion of flagella and cilia to reproduce, and that very motion probably came from some spirochete. The motion of our sexual reproduction is the motion of syphilis. Spirochetes, relatives of those that give us this especially terrible venereal disease once may have clung to the outside of a protist. And, as *Mixotricha's* spirochetes did, they stayed forever, not only a part of the cell but the motion of the cell. Donating their particular motion in order to help us reproduce could be either a foul scheme for widening the spectrum of the spirochete curse, syphilis, to include an unborn child or perhaps a belated atonement for the death and anguish caused by its venereal crimes, acting as midwife for the miracle of conception.

In both human males and females, the cilia make

reparation again for their ancestral crimes. They line the entire surface of our respiratory tract and by their perpetual motion, spread a coat of protective mucus and beat away any foreign particle wandering in with the inhaled air. Cilia are probably our most effective anti-lung cancer agents, a fact that cigarette manufacturers do not know or do not want to know. Cigarette smoke paralyzes the cilia, completely defeating several million years of evolution. Tars in cigarette smoke, asbestos fibers, or any cancer-causing particles, are free then to move in and settle disastrously in a cigarette smoker's lungs.

What I like most about these hard-working fibers is that they may have been the origin of our humanity; the fibers invented by the prokaryotes may have, in turn, invented our brain. We have to go back to *Euglena* to find out how our brain, where consciousness and humanity reside, begins. The flagella-propelled *Euglena* has a bright orange, light-sensitive eyespot. The angle at which the light hits the eyespot, directs the angle at which the flagellum will be held while it beats. Light intensity also adjusts this flagellar rudder. The flagellum connects directly to the eyespot and *Euglena* can direct itself, because of this relationship, to where there is bright (but not too bright) light. This is without a doubt, perception in the protists, and response to what is perceived. The *Euglena*—or rather the flagella of the *Euglena*—respond to the interpretation of the environment by the eyespot. This is on an incredibly simple level, but thought and comprehension had to start somewhere and there is good evidence that it was here, in the spiraling nine of the flagella. The spiral nine is a common structural foundation for sensitive sensory receptors: the chemical detectors in insects, the touch receptors on the suckers of an octopus, the light sensors at the end of each starfish arm, the blue eyes of the scallop that peer at us from between the crenellated shells. We use this famous spiral well: for the gravity detectors in our ears, for our sense of smell, and for the light-detecting visual cells, the rods, in the retinas of our eyes.

The simplest nervous system needs just a sensor (to receive the information from the environment) and an effector (to act upon it), like an eyespot and a flagella. In our visual cells, the transmission of the message to the optic nerve is handled by a bundle of—cilia. The same structure as the flagellum that transmits the message from the eyespot along its length, translating it into motion as it goes. Motion, of course, came first. Then vision. An eye is merely an adaptation for making motion more efficient, more meaningful. Vision is really inactive locomotion; it lets the organism learn the same things about the environment as it could learn by moving, but it saves a great deal of energy.

There were eyes of many kinds long before there was any kind of brain. Simple eyes at first, that didn't need a brain. All they needed was a way of responding, a simple effector. We have always learned that the eyes are direct extensions of the brain, in direct physical contact with it through an inch or two of optic nerve. It makes more sense the other way around. The brain is not only an extension of the eye, but an invention of the eye. Eyes did not remain simple. As they became more and more complex, they needed a way of interpreting the complex images they were receiving. And so, perhaps they built one, using what was at hand, odd flagella or cilia. For the protein found in the axons of the nerve cells—their delicate, connection-seeking, message-transmitting, cell extensions—is tubulin. Tubulin is the very protein of the sliding filaments in the spiral nine of flagella and cilia.

FIVE

Genes

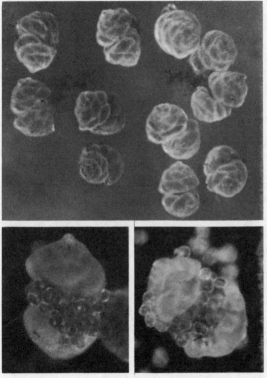

(courtesy of Prof. Karl Grell)

Plate VII. *Glabratella sulcata*, wearing their glass turbans while mating.

L ove can be very complicated. It can be sighing, sonnets, and suicide threats; songs, sweat, and tears. But whatever the plummeting depths, the unbounded joys, the countable and unaccountable ways of love are merely adjuncts to a system, the sole end of which is to fuse two cell nuclei together. That's all. One nucleus comes from the male and has half of his genetic information; the other nucleus has half the female's. Their fusion is sexual reproduction, the *raison d'être* of love. Sex was quite an invention. Once it was invented, it was kept and adopted by almost every living thing because progress—at least as far as evolution is concerned—requires variety. Natural selection needs an omnifarious, perpetual supply of new organisms to experiment on and to select from. There is a constant search for a better way, a more efficient use of environmental assets. This means new combinations of genes must be made and new environments for the genes must be tried; and for this genes from two different organisms are brought together in a single cell, or a single new organism. Love may be very complicated, but sex is really quite simple, as you see.

And it is fair. Whatever happens afterward, at the very beginning of life, male and female play exactly equal parts; for every heritable trait there will be a pair of genes, one that originally belonged to the male, the other to the female. The genes themselves are triumphs of molecular design; they have perhaps the most difficult task to perform of all biological molecules, one that involves simultaneously maintaining two opposite and contradictory properties. They have to keep an incredibly delicate balance between change and constancy, mutability and immutability. For evolution to make any sense, evolutionarily successful parents must

89

have successful offspring; genes have to be steadfast enough to guarantee this. But they have to combine their constancy with capriciousness. They have to be able to change often enough that genetic experiments can be made; so that if environments change there will always be someone, somewhere, who can function in the new environment. Evidently, dinosaurs were low on genetic variability. They had the right genes to rule the earth, unchallenged, for about 100 million years. Then a great change came into their lives. We don't know what the change was; it certainly did get colder. Whatever it was, most dinosaur genes could not meet the challenge. Among all of the dinosaurs that walked, ran, or swam, there was not a single one that could survive. They had to pay the price of being genetically conservative. Today there are no more dinosaurs. However, there may have been a small pocket of dinosaur DNA that allowed some of them to fly from the dreadful fate of their more earthbound relations. Dinosaur genes fill our skies today. All the essential characteristics of birds —wings, scales, feathers and warm-bloodedness are legacies from their dinosaur forebears. When we consider that there are fifty birds for each human being, we may not be able to call the dinosaurs true evolutionary failures.

Only one kind of molecule was able to manage the paradox, the incredible juggling act of evolution: constant stability and constant instability. These were the nucleic acids, and they won their success early. Every cell of every organism, with the boring predictability that characterizes evolutionary success, makes its genes of deoxyribonucleic acid—the DNA of Nobel Prize and best-seller-list fame. At the level of the genes, all creatures great and small, simple and staggeringly complex, are equal without distinction. All life lives and repro-

duces by information coded in molecules of DNA, organized into the genes. The paradoxical genes, with a
duality that generates two confusing but equally true
tenets of biology: all living things look like their parents
and all living things do not look like their parents.

Nothing is more eloquent than a gene at pleading the
case that natural selection is an energy-stingy, material-
thrifty, efficient-expert force. The form of a gene is
the essence of its function, in breathtaking perfection.
The genes are strung together in the precise, serpentine
coils of DNA, two strands spiraling around each other.
The molecular structure of this seductive double spiral
is almost too simple to believe. No matter how long a
DNA molecule is, it is always made up of only four
subunits; there are four "bases": adenine, guanine,
thymine, and cytosine (A, G, T, and C). On each of the
two strands, the bases are lined up, one next to another
along a backbone of a chain of sugar molecules. Each
base is attached to one sugar and stretches invisible
bonds over to another base wound into the opposite
strand, forming the steps of this graceful, mysterious
spiral staircase. A, G, T, and C are the only letters
in the genetic alphabet, yet they can spell everything
that lives. Actually, it's more accurate to say they can
be translated into everything. The genes are cryptic.
DNA is really a code, not the actual information that
runs the cell; it has to be translated before it means anything to the cell. All genes (except for a few) are finally
translated into enzymes, the regulators and directors of
the cell. Like some sacred text that is intelligible to only
a few, yet is used to govern a complete society, the
genes require intermediaries between themselves and
the particular society they govern.

To make a human being requires spelling out about
100,000 enzymes, quite a task using only A, G, T, and

C. It is not such an impossible task when we know that the code is really translated into an alphabet of twenty letters; each enzyme is a protein, and proteins are built from combinations of twenty different subunits, the amino acids. If we have an alphabet of twenty letters, it is easy to form a language as complex as English, Russian, or Turkish. Each of which have a vocabulary of at least 100,000 words. The complete language of life—including the genetic code—is far easier to master than any of the languages of man.

The relationship between DNA and the amino acid alphabet is clear and precise. A sequence of three bases in the DNA strand makes one code unit; the genetic code is a "triplet" code. With the four letters available, we can make sixty-four different triplets (AAA, ATC, GCA, and so on). Each of these sixty-four triplets means only one thing to a cell, but the message has to be decoded first. Each triplet stands for one particular amino acid. A gene contains the complete instructions for putting together an enzyme or a protein: a chain of amino acids linked together one by one. One gene equals one amino acid chain. That is the beginning and the end. But it is the decoding machinery between the two that is truly a wonder of any world, ancient or modern. I cannot believe that even a conscious design by an architect—human or human-directed computer— could surpass this one for the sheer simplicity and elegance of its plan and the variety of its product.

The genes always remain cloistered in the nucleus, protected as much as possible from the outside world, these molecular medieval monks have the task of copying endlessly the wisdom of the world and preserving it for those able to read it. All our cells (and our selves) will ever need to know is behind those thin, diaphanous walls of the nucleus. Before our cells can actually comprehend the coded scrivenings, the information has to be carried beyond the double wall; for proteins are not made within the nucleus but outside, in the cytoplasm of the cell. The intermediary, moving between the genes and the machinery of the cell, is messenger RNA (a ribonucleic acid, very much like DNA, but with an oxygen in the backbone sugar that DNA does not have). The messenger molecule itself is a copy of the genetic message, still in code. The transfer of the code is made very easily, for the bases of DNA are highly reactive. Molecularly, a DNA base cannot bear to be on its own and forms, as soon as a partner is available, spontaneous bonds with another base, an atomic handclasp that makes DNA the bifurcated little devil we all know it to be: one strand of the molecule joined to the other by these pairing bases.

This intense need for companionship of the DNA bases is remarkably fastidious. A pairs only with T, and G with C. One strand of DNA is not identical to the other, it is complementary. Each strand is a complete but not exact reflection of the other, like Tweedledum and Tweedledee. If we know the order of the bases in one strand, we know the order in the other, we just have to interchange A and T, G and C. The spontaneous passion for pairing guarantees that messenger will carry an exact copy of the genetic information. RNA bases will pair with DNA bases in the same complementary fashion, except that in the RNA alphabet there is no T. There is a U (uracil) instead, but U pairs with A, exactly as T does. When the time comes for the gene to be read, the two strands of the DNA double helix separate, and RNA bases pair with the bases in DNA strand. An enzyme makes these RNA bases into a single long strand; where there was A in the DNA, there is now U in the messenger. Where there was a T, there is an A, for a C there is a G, for a G there is a C. If the DNA had read AGC, the messenger carries the message as UCG. The information is still in code, still a triplet code, each triplet a *codon* in the messenger. The messenger is made into the message. The messenger *is* the message. Once the gene has been transcribed, the messenger leaves the nucleus for the protein-making machinery—tiny round structures, the ribosomes. We have come to the last step, the translation of the genetic code into a protein.

Ribosomes are mediators; they bring the messenger together with the interpreter, molecules that finally translate the code, transfer RNA. If these translating molecules (the transfer RNA) were pressed flat in two dimensions, they would look like three-leaf clovers with long stems. (In its normal three-dimensional shape, a transfer RNA molecule assumes a Daliesque shape, twisted and bent, like a limp watch or really just like one of those madly ubiquitous and apparently self-reproducing wire coathangers after wrath has been vented upon it.) Whether or not we look with a more or less artistic eye, the translating ability of transfer RNA lies in two separate parts of the molecule; one part reads the message in the codons of the messenger and the other translates it into something the cell can understand, a protein chain.

Transfer RNA, like messenger, is built of subunit bases, and at the top of the middle leaf are three special bases, an anticodon. The anticodon triplet will bind to a complementary codon in the messenger. A UUU triplet in the messenger (which was AAA in DNA) will be recognized as an appropriate pairing partner only by one transfer RNA, the one whose anticodon reads AAA. And each different anticodon tells us that, tied like a pennant to the stem of the clover, is one particular amino acid. Messenger comes to a ribosome, and so do transfer RNAs, each one flying its own unique amino acid. Two at a time, transfer RNAs join the messenger at the ribosome, attaching to it by their anticodons. The amino acid carried by one transfer RNA is linked to the amino acid on the other. The transfer RNA falls off the ribosome when it has lost its amino acid this way and goes off somewhere in the cell to get another. Another transfer RNA, with its own amino acid, steps onto the ribosome, but only the one whose anticodon matches the codon next in line on the messenger. The chain of two amino acids is now linked to the amino acid carried by the new transfer RNA. Another transfer RNA falls off, another steps on and so the protein chain is formed. Two amino acids, then three, then four. But each amino acid is put into the chain specifically at the instruction of the messenger. One at a time, following these instructions, amino acids are linked together.

There are sixty-four possible codons that the messenger can spell and each amino acid has its own codon or set of codons. Some amino acids are coded by several different codons, some by only one. And this is how the only truly universal cellular mechanism works. The code in the genes is transferred, in complementary form, into code in the messenger. Transfer RNA molecules simultaneously read the coded message and translate it, codon by codon, amino acid by amino acid, into a protein—an enzyme that will direct the cell in a specific task. So, you see, it really is a far easier language to learn than English or Russian; an Esperanto understood and spoken by any cell. The genetic code is *the* code, the only code. Every cell operates by genes and each gene gives its directions in the same code that is translated through messenger, ribosomes, and transfer RNA. And the parts of the machinery, too, are universally interchangeable, from man to mouse to microbe. Bacterial ribosomes will make proteins from mouse messenger, and mouse cells can translate human DNA just as easily. As Tennyson once said about something totally unrelated, "Most can raise the flowers now, for all have got the seed." We can now grow exotic genetic flowers, for we have the seed of the genetic code.

Virus can raise the flowers. They can replicate their own genes at will even though they do not have their own translation machinery, because the genetic machinery has so little individuality, there is nothing about a strand of DNA that can tell a cell it is a stranger. Even if the DNA will be fatal to the cell and to its organism. This is why viruses are the ultimately successful cellular guerrillas. They are not even alive, merely some genes wrapped in a coat of protein and yet, because of the universality of the genetic machinery, we must suffer from their plagues, their deadly flowers. Even our superb intelligence has not freed us from the depredations of these ridiculously simple beings. A virus has to have only one talent to be completely successful: it has to attach to a living cell and inject its genes. It merely pirates the rest of what is needed to reproduce itself, sometimes a thousand times over. Once the viral genes are inside the host cell, they liberate the translation machinery of their host for their own nefarious purposes.

Sometimes they allow the host a limited use of the machinery after the liberation; the infected cell is able to carry on but often, just as in the military occupation of a country, all of the resources of the victim are put at the invader's disposal. The viral genes simply take the cell over; only viral DNA is made, only viral genes are decoded into viral proteins, but the cell supplies everything that is needed, at its own expense. Within

an incredibly short time—sometimes only an hour—
there is a complete new army mobilized, one hundred
to one thousand new viruses, equipped with all they
need: their genes. To get out of their host, they destroy
it, cold-bloodedly using the cell's own enzymes; the
viruses persuade the cell to commit suicide. Each of
these new viruses can start a new round of infection,
one virus per cell. In another hour, each of these infected cells is destroyed, having been ruthlessly used
to manufacture another troop of viruses. This is nothing
less than real warfare; it is invasion, amoral and in-
humane, and does not follow the Geneva Convention.
It's a marvelous technique, though; we have seen it
work in polio, hepatitis, influenza, encephalitis, and
Vietnam.

And our usual techniques of combating disease work
about as well as conventional military techniques work
against guerrilla warfare. We cannot kill the enemy,
for the enemy is an integral part of the last thing we
would want to destroy; our own selves. Our body has
its own defense system, one that can destroy virus, but
it takes about a week for it to be mobilized. And so
every virus infection turns into a race: can the body
build its defenses and destroy the virus before the virus
destroys it? It is not a race always won by the body.

As if we don't have enough to worry about with
viruses using our own cellular seeds to grow their
sometimes deadly flowerings, we also have to worry
about the consequences that scientists in their labora-
tories have the seeds of life as well. What kind of flowers
are they going to grow in their gardens? We know the
relationships between all parts of the eternal and infernal gene machine. This has turned us into Prosperos
of the cell, for the knowledge we have pried out of the
cell, molecule by molecule, has given us the beginnings

of an almost unholy power. Just as the discovery of another marvelously simple relationship, $E = mc^2$, gave
us a dreadful power over the atom. Up until now, there
has been a certain innocent exuberance about the experiments we have been trying; we haven't quite gotten
over the intoxication of finally seeing the bones of
beauty bare. Some experiments were done just for the
sheer delight of persuading ourselves we finally did
know the secrets that before the last quarter-century
were considered almost unknowable. There was something of an innocent childish revel about it all. But now.
Now our mastery appears so complete that some find
it frightening, more frightening than rabies or encephalitis. We can now make an artificial, but completely
workable gene, gluing it together base by base, merely
by translating backward from the order of amino acids
in a protein. We can make a fake gene—one that has
never seen a cell before, and apparently only *we* know
for sure—the cell can't tell the difference. For this particular artificial flower of genetic experimentation, the
pseudogene was injected into a bacterial cell, and the
bacterium was completely taken in by the deception. It
put the gene into its own DNA and read it as faithfully
as if the impostor DNA had been its very own; the protein was made and used by the cell. What hath been
wrought by our curiosity and our opposable thumbs?

Well, that depends upon to whom you are listening.
This lovely experiment that took nine long years
brought smiles of pleasure and satisfaction to many
molecular geneticists. How reassuring. What we had
thought was true is actually true. The exact relationships between gene and protein were beyond question
now. How perfectly horrible, others cried, waving the
specter of genetic engineering in front of us, again. As
usual. By now this specter should be so transparent and

motheaten that it wouldn't frighten a child. It probably doesn't. But some adults always can be counted on to cry "Frankenstein" and panic.

We can also count on the journalists to wave the sad specter in the hope that they will panic their audience. Journalists who know as much about cell biology as "a hog knows of predestination," to borrow from Mencken. Now that we can make any gene we want, trumpet the ever-watchful watchdogs of the media, we will be able to mold innocent children into Hitlers or Mozarts at will (or, at least mold an innocent fertilized egg). Well, we can certainly make a gene. But not any gene we want. For that we have to know the exact structure, codon by codon, or amino acid by amino acid, of the messenger involved or the final gene product. And you know, it just ain't easy. Except for our red blood cells that make only hemoglobin, the cells in our bodies are making tens of thousands of different proteins, decoding the same number of genes into messengers, all at the same time. A complete, indecipherable mare's nest. Journalists will propose that we should be filled with fear and trembling—not to mention loathing—at each indication of progress in modern genetics. They never propose just how we should go about finding the gene we want and getting it out. That at the moment is completely and utterly impossible. Even if we assume we only have 100,000 genes, so far we have identified fewer than 3,000 of them. Knowing a gene exists doesn't necessarily mean we know what the product is. For many of them, including the ones that cause muscular dystrophy and Huntington's chorea, we simply do not know what the gene codes are. Nor do we know how to find these genes. We can locate only about 200 of those 100,000 genes, placing them on one of the twenty-three pairs of chromosomes that hold our genes. We haven't been able to pinpoint them too precisely; we can just say that they are somewhere near one or the other ends of the chromosome or somewhere near the middle. One cannot just pick the human gene of choice out of the cell, like a plum from a fruit basket.

But supposing we do have the gene we want—we have made it, or stolen it, or borrowed it. What then? Not a whole lot more. The artificial gene we have been talking about was put into a prokaryote without a nucleus, and prokaryotes are notoriously sloppy about handling genetic information. Almost any piece of DNA that gets into the cell will be adopted. Bacteria are really quite charitable and easygoing about their genes. They can afford to be. It is no great loss if a bacterial cell picks up and uses a gene that is completely wrong for it. Wait half an hour and there will be another complete bacterial generation. Mistakes are cheap when the organism can be replaced that quickly, but they get more expensive the longer it takes to replace the organism. For this reason, eukaryotes—especially multi-cellular ones—take good care of their genes, protecting them by a double nuclear wall, and concealing them in the incomprehensible tangles of the chromosomes.

In spite of these barriers, impassable at the moment, there will be a day when we can get the gene of our choice into a human nucleus. I wish it were today. Our paradoxical genes bring us a lot of grief. Molecules are only human; they make mistakes. The structure of a gene can be changed spontaneously. These mistakes are supposed to happen, to feed the insatiable appetite of natural selection for new ideas, always more new ideas. But even a tiny molecular slip—the change of a single base in the DNA—can be fatal. The translating machinery is not programmed to detect mistakes, and we suffer for it. There are several thousand genetic diseases, some quite horrible. A gene has the wrong information; the protein comes out wrong, misshapen, and it cannot do what it is supposed to. Most of these genetic diseases cannot be cured; many are fatal. Many aren't fatal, but it would be kinder if they were. It hardly seems fair that our genes are an experimental playground for such a ruthless, impersonal force. Hardly any of the misspellings ever make a new kind of sense. Each organism alive has been honed into perfection by millions of years of evolution; the interrelationships of the systems are delicate and elaborate, and can be permanently disrupted by one single wrong base out of the three billion pairs we have. It seems a hard life this one, what with the competition for survival, millions of unsuccessful genetic experiments (one of which might be a child of ours), just so one experiment might work. But that's progress. And the experimentation continues. Evolution's and ours. The aim of ours is to counteract the effects of evolution's. To one day be able to cure genetic diseases, to repair genes gone wrong. That is a consummation devoutly to be wished.

What of those unscrupulous dictators we are told about, the ones that will persuade some equally unscrupulous scientist to put together the genes that are necessary to make the citizenry they want; super-smart citizens or super-dumb, whichever will suit their nefarious schemes. I can't imagine anyone considering this an intelligent plan, just what does one do with 10,000 helpless, diaper-requiring babies, no matter what kind. (This is assuming that one of these unscrupulous dictators has managed to round up 10,000 willing wombs to incubate his genetic experiments.) Every day we see that it is far simpler and surer just to propagandize, and get the same results as tedious, messy and technically improbable experiments; all Mao needed to accomplish this was a little book bound in red plastic.

We really have so little idea of what makes a human being. We know only that it is

> a self-balancing 28-jointed adapter-based biped; an electrochemical reduction-plant, integral with segregated storages of special energy extracts in storage batteries, for subsequent activation of thousands of hydraulic and pneumatic pumps, with motors attached; 62,000 miles of capillaries; millions of warning signals, railroad and conveyor systems; crushers and cranes (of which the arms are magnificent 23-jointed affairs with self-surfacing and lubricating systems, and a universally distributed telephone system (needing no service for 70 years if well-managed); the whole extraordinarily complex mechanism guided with complete precision from a turret in which are located telescopic and microscopic self-registering and recording range-finders, a spectroscope, *et cetera,* the turret control being closely allied with an air-conditioning intake-and-exhaust and a main fuel intake.
>
> Within the few cubic inches housing the turret mechanisms, there is room also for two sound-wave and sound-direction-finder recording diaphragms, a filing and instant reference system, and an expertly devised analytical laboratory large enough not only to contain minute records of every last and continual event of up to 70 years experience or more, but to extend, by computation and abstract fabrication, this experience with relative accuracy into all corners of the observed universe. There is, also, a forecasting and tactical plotting department for the reduction of future

102

possibilities and probabilities to general successful specific choice.*

And the knowledge of how to build this machine requires at least 100,000 genes, and perhaps as many as
5,000,000. We have enough DNA to make 5,000,000
genes. To protect these, there is a relatively tamper-
proof system, the cell. So far it has done its job well,
at least for these last few billion years, until we started
messing around with test tubes and electronic equipment. It takes far more than just the genes to make a
human being and more than the nine months incubation
in a complex environment constantly in a state of nutritional, hormonal, and heaven-knows-what-else flux.
We are profoundly ignorant as to why one Austrian
turns into Hitler and another into Mozart. We know a
lot about the genes, it's true, but we don't know enough
to make human beings to order. And we won't know it
soon enough for me. Or for the 20 percent of the children in pediatric hospitals because of defective genes.
If you want something to worry about, may I suggest,
rather than custom-made human beings, that you worry
about the millions of human beings who are going to
starve to death, or the millions that will be poisoned
by or get cancer from our industrial trash, black licorice,
or red maraschino cherries *before* we can successfully
push human genes around. If you lie awake at night
worrying that someday, somewhere, someone will dic-
tate the genetic messages of a human being, sleep well.

*This description is taken from the Buckminster Fuller book,
Nine Chains to the Moon and is, I find, a precise and thorough
definition of the physical human being, a magnificent machine,
not including the characteristics bestowed on it by the non-
physical qualities of mind, spirit, soul, whatever.

You'll probably die in an explosion of your local nuclear power plant before that happens.

A meter of DNA makes a human being. That is enough for 5,000,000 genes. At the most, a human cell needs a few hundred thousand. What are our cells doing with the rest of those genes? What are the rest of those genes? Where do they come from and where are they going? We know that we have several copies of certain genes, probably the most vital ones. These extra copies could be a fail-safe mechanism, so that evolutionary genetic experiments could continue with less chance of killing the organism by misspelling a crucial gene. But even taking into account these extra gene copies, we use only a puzzlingly small fraction of our genetic material (from 2 to 50 percent, depending on the estimator) to stay alive and function as a human being. Not many biologists have any suggestions about what all of that extra genetic material is doing, and why it is valuable enough for a cell to carry around, even if it doesn't use it. My best friend, who is not a biologist, has a suggestion. He is sure that we are carrying around time machines in our nuclei. Although we don't use the extra genes now, we used them once. And we might use them again. Most of the genes that we had when we were stooped, inarticulate man-apes, several million years ago, are still with us. Probably even some of those that were with us tens or hundreds of millions of years ago. Our pedigree, as we went from reptile to mammal to primate ape, is recorded in our nuclei; it is our evolutionary history written in genetic code. But it is turned off, repressed, as we know that genes not in use often are. Liver cells have all of their genes turned off except the ones they need to make them good liver cells. And in our red blood cells, almost every gene is turned off except for the few necessary to manufacture hemo-globin.

Our repressed mammalian genetic heritage does keep resurfacing here and there. One gene, for supernumer-ary nipples (double rows of multiple nipples) that most of us left behind when the primates branched out of the mammalian family tree, reappears quite often, to remind us we once had litters of offspring to care for instead of just a pampered one or two. At least one in every two hundred people is born with one or more extra nipples. (An extra nipple is said to have been a part of the charm Anne Boleyn held for Henry, and in the Middle Ages it was considered positive identifi-cation of a witch, keeping the burning stakes well sup-plied with victims.) We are also occasionally born with vestigial tails, a few extra vertebrae at the end of our spines. These ancestral tail genes are still there, as are those for tail-wagging muscles (we all have those, I understand). The excess genetic baggage we carry around, may just be records of our reptilian years, our mammalian years, and our primate years, but re-pressed, turned off because we don't need that informa-tion any more.

We know our genes don't change very rapidly. It is more than 30,000,000 years since we bid genetic adieu to the hairy apelike ancestor we shared with the orang-utan and the gorilla; and 15,000,000 years has passed since we started off along our own peculiar evolution-ary trail and left the chimpanzee to take its own turn-ing. Yet in all that time and with all those apparent changes, our genes have themselves changed very little. Our genes are only one percent different from a chim-panzee's. All of the other sequences of DNA that make our genes we share with that amusing but disturbing ape, *Pan troglodytus.* The difference between a human being and a chimpanzee lies within that relatively meager portion of DNA that we don't share. We have exactly the same genes for hemoglobin as chimpanzees do, those molecules don't differ by so much as an amino acid, so our human hemoglobin genes cannot be more than a DNA base different from chimpanzee hemoglobin genes. We are two amino acids different from the gorilla in terms of hemoglobin, even though we left them behind at least 30,000,000 years ago. Our blood types—A, B, O and AB—are shared with chimpanzees, orangutans and gibbons, but not with baboons or pig-tailed macaques, which should be comforting. (Gorillas all have type B blood.) So, only one percent of our genes seem to be uniquely our own. And of the remainder, that we share with our fellow primates, what percentage is uniquely ours as primates, picked up since we evolved from the tree-shrew, sharp of tooth and temper, who gave us all our thumbs and brains? How far back does our time machine go? What kind of traits would appear if we could turn on again these turned-off genes?

My friend thinks we might still have the genes to fly. Those reptilian archosaur genes that taught the vanished pterodactyl to fly and gave wings to the eminently successful birds. These archosaur genes may even have given flight information to our cousins, the bats. Will we end up flying? After all, flight seems to be the end point of evolution; most insect species fly, and those flying reptiles, the birds, fill our skies; among mammals, only rodents outnumber the flying bats. A schoolchild joke comes perilously close to being a fundamental evolutionary truth. "Why do birds fly south in the winter?" "It's faster than walking." It is also much cheaper. It takes less energy than walking and it certainly makes the organism more efficient as a food gatherer, with three dimensions to work in rather than two. It may be that we can turn our nuclear time machine on in the future and learn to fly; among the at least 800,000 unused genes we have, there may be flight. Birds and mammals came from the same reptilian ancestor somewhere along the way. Bats may have just renewed the flight capacity; they just turned on their archosaur genes that gave them light bones, small size, and weird fingers. The Icarus myth that fascinates us so may really be in our genes, not just in the imagination of an ancient storyteller. Flight is beloved and envied by almost every human being, an integral part of our myths, dreams and religions. Our flight fantasies could just be our old archosaur genes, repressed at the moment, but still nudging our unconscious, nibbling barely at our consciousness.

Our brains are getting smaller, a significant fact. As we evolved from the apes, our brains ballooned. Each progressive step in evolution is marked by skulls of increasing size. As Neanderthal gave way to Cro-Magnon and we ascended in wisdom, so our foreheads ascended and the tops of our skulls became farther and farther from our noses. But now, our brain, which had increased fourfold in size from that of the first hairy ancestor of true humans, is no longer growing. The expansion of our forebrain, one of the few really unique human developments has come to a sudden halt. Over the last 100,000 years, we have made the transition from a semistarved hunting and hunted animal to the machine man of the nuclear age free of any increase in brain size. In fact our brains are slightly smaller than the average Cro-Magnon's.

Of course, this just may be because we don't need all that brain anymore. Superiority of mental process,

reflected in tool use or technology, is no longer re-
warded with reproductive success, the way it was at the
dawn of time. It is not hard to imagine that the human
being who could build a better spear or who could remember and even predict where game could be found
would leave more children (and thus more genetic
copies in the next generation). Not so today. All members of a society can benefit from the intellectual advances made by only a few in that society. This is culture. It is a transmission of information and a willingness
to share (or if not to share, at least a willingness
to sell) information and advances with all members.
All members can benefit from the mental processes of
a few and do not suffer reproductively if they are not
innovators.

We don't need all of our brain today. No feat greater
than that of remembering the location of the supermarket is necessary to ensure that our children have
food, and perhaps also some degree of manual dexterity
to free that food from the impenetrable clutches of
modern packaging. We also note a widespread and
ever-increasing disinterest—even an active antipathy—
for the processes of gathering information and making
our own deductions. Our major sources of information
more and more do not just present facts; they present
ready-made conclusions and opinions for us to adopt
in toto. News magazines and news commentators make
substantial contributions to conditions where our nerve
cells can lie in rust and disuse like abandoned railroad
tracks. Even complex problems need only the combination of a computer and a computer programmer
who can be relied upon to free us from many of the
burdens that the process of thought might impose.

Or maybe I just have cause and effect reversed. This
could just be adaptation in preparation for flight, not

the effects of disuse. The final evolutionary step toward which our dreams are driving us. Darwin said that the fittest survive, and it seems that the fittest fly. Our brains are shrinking (toward the size of a bird's?) and so are we. Children are no longer growing to be taller than their parents. It just might be our archosaur genes switching on.

SIX

Communication

We are all engaged in constant, clear, unambiguous conversation. Not with each other, of course. It is our interior cellular dialogue that is so articulate. Communications between one human being and another, if we believe our various media mentors and paperback pundits, are conspicuous for their opacity and duplicity. It may be that communicating intelligibly with each other is one cellular accomplishment that does not necessarily become property of the organism. On the other hand, we may very well be able to say what we want, it's just that we don't know we want to say it, or we don't know what it is we want to say. It certainly wasn't that way in pre-Babylonian days, before cells got organized into human beings.

The first attempts at communication probably came along with sex; at least there could not have been sex without it. But this was simply recognition of mutual attraction and need. When protists started forming societies, banding together to be more mobile and less edible, communication had to go further than the "Me Tarzan—You Jane" acknowledgment that would have been the minimal requirement for sex. The protists exhibit their usual extravagant empiricism in their social planning. Today they present us with a full, living spectrum, from casual slapdash congregation with little necessity for communication, to stratified societies whose communication systems are still being used by multicellular organisms.

Actively hunting predators, whether human, feline or protist, are not particularly good candidates for stationary, stable social interactions. Amoebas or Paramecium, that must move around to find their food, would be hampered, not helped, by any societal obligations that required them to stay in one place. It was photosynthesis that allowed protist societies as, in the form of crops, it allowed man to participate in a new level of organization. We find permanent protist societies only among those species of green protists that make their own food, or those that eat by absorbing the dissolved food in .their watery surroundings. *Euglena,* usually independent, free-moving and flagella-flipping, sacrifices its independence and flagella, but only occasionally, to form a temporary and almost completely noninteractive society. When a *Euglena* is ready to reproduce asexually through simple division, it discards its flagella and surrounds itself with a transparent jelly nimbus. Protected from harsh sunlight and from certain predators in this gluey sphere, *Euglena* are free to give their full attention to other matters. They divide several times, each daughter cell remaining within the small gelid world its parent created. There is no communication or social intercourse between parent and child, no recognition that they are all the green fruit of the same green loins. And, once division is done, all of the cells furnish themselves with flagella once again and swim away without so much as a backward glance of their red-orange eyespots. But we can hardly call this a real society; it is more a temporary noncommunicating family. The first true society with cooperation, and with actual physical and communicating links between the members is *Gonium:* green, flagellated, and poor group planners.

Gonium, although its members have agreed on a common goal, ends up like a disorganized commune or one made up of anarchists. They come together as a slightly rounded disc, like a shield, all protists in the *Gonium* colony shoulder to shoulder, heads forward, flagella straight out and perpendicular to the disc. By flapping their flagella straight out behind them, they guarantee themselves the hardest work and the least progress; for pushing their communal disc forward through the water, broad side forward, they are choosing the line of greatest water resistance. They either are very poor at hydrodynamic theory or each member demanded the right to be first. *Gonium* progress is slow and inefficient, but it is a pure democracy; each member of the society leads the parade. The members of the *Gonium* colony are connected to each other by thin protoplasmic threads, so they could be coordinated to propel themselves in a more efficient fashion, but they have chosen equality.

Pandorina, on the other hand, has renounced democracy; even though the member cells are equally spaced in a ball, their formation belies the organization of this society. Here we find the first hints of what we might call an Ayn Rand philosophy: it is agreed that only some can be leaders, the others must be followers. You can tell the leaders in *Pandorina;* they are at the front, as the ball moves through the water, and as emblems of their exalted positions, they are granted larger red eyespots than the "followers." The privileged among *Pleodorina* citizenry are not so gentle in all their expression; only some of the cells in the colony are allowed to reproduce, to attempt immortality through the passage of their genes. Something is passed through the colony, through the frail transparent struts of protoplasm, that forbids the anterior cells of the colony to reproduce. Their genes and their bodies have mortality imposed. There is no reprieve. If one of these cells leaves the colony, it will die, just as the colony will die if the connections between the cells are severed. The anterior cells must stay and await their fate; sexless workers whose job is to contribute mass and movement—but not genes—to the survival of *Pleodorina.*

The end of this series—the attempts of photosynthe-sizing, flagellated cells to organize—is *Volvox*, a soci-etal master and an incredibly beautiful creature. I can describe the society easily, and will, but the only way I can describe *Volvox* is to say that its beauty is in-describable. It is a gleaming, shimmering bejeweled sphere, studded with transparent green cells, rotating and luminous with reflected light. Each cell in the colony is threaded to six others by fragile protoplas-mic skeins, a communications web whose instructions will always be obeyed. A colony of *Volvox* can contain from 500 to 500,000 separate cells; a village or a city. Spun by the coordinated beat of flagella, *Volvox* moves, each cell a fixed star in a revolving, transparent firma-ment, each cell helping to move the colony into its place in the sunlight. If the spot is too dark or too light, there is an extra flagellar kick from the cells on one side, spinning *Volvox* to more ideal conditions. It isn't visually apparent, but a *Volvox* colony contains a very small select elite. From the cells in the southern hemisphere a few are chosen (just two or at the most, fifty) to carry on, to form new colonies of their own. These can reproduce either asexually, by merely di-viding, or sexually. The reproductive cells can pro-duce male or female gametes that will fuse to become the new generation. The cells that will just divide to build another spectacular geodesic dome do not ever get to experience sex, and so they are deprived of an experience that could have been exciting for them. Nonetheless, they can manage a spectacular trick that is terribly exciting to us. These particular reproductive cells divide and form a new but unexpressed colony on the inside of the parental sphere; all of these new cells are flagella-less and facing inward. Then, when their time is approaching, they turn themselves inside

116

out through a pore in the wall of their colonial parent, a balloon blown inside out. When they find themselves outside, their heads now pointing outward, they grow flagella and swim away, a brave new colonial world. Since new colonies are not released until the old colony is ready to die, these daughter colonies are sometimes kept at home long enough for them to have daughter colonies of their own, hanging baubles, awaiting the distintegration and death of their progenitive society. Only then will they be free to live on their own.

The southern cells that develop into male gametes with the motion and function of sperm, can perform the same topological trick. Once they are free, they separate and swim off to find a female cell which is still a captive in the colonial web. Only a fertilized female cell can leave home and grow a hard, spiny coat for protection against the cold winds of winter. This new colony doesn't swim, but falls to the bottom of the pond, to rest there inanimately until the spring. Then it joins in with the general verdancy, shedding its dowdy, practical brown coat, to blossom into spinning viridescence. A complete, well-balanced world has come to life again, to orbit through its own watery universe.

The family of the art nouveau Tiffany creation, *Vorticella*, is apparently devoted to visual aesthetics. One close relation, a colonial, is not content even with the perfection of the geodesic dome of *Volvox*. It has to flower into an ornately elegant tree, from which the component cells hang, like the florets in a lily-of-the-valley. The trunk and the branches of the tree are constructed by these tiny creatures, a community effort, but they are not parts of the cells, merely serving as conduits for the protoplasmic communication system that runs from cell to cell. A new colony starts when one of the protists leaves the family tree and sets out

117

on its own. For a few hours it is free, the only freedom it will ever know in its life. Then it finds a place to settle, quite often the back of a turtle shell. It can tell the difference between a good homestead and a bad one, for the migrant cell will sometimes try several sites before it decides to begin construction.

The first step is to build a stalk, and then perched on this, the pioneer divides; one of the daughter cells stays at the tip, the other is displaced a little to the side. Then the tree begins. Each of these cells divides again, one daughter always remaining at the tip, the other sitting aside. The tip cells are the builders, the ones that continue to make the stalk and the branches of the colonial tree. The cells on the side do nothing but are purely decorative; their only task is to adorn the branches. Each tip cell divides, builds a short section of stalk, plugging into the general communications cable, then divides again. And thus they grow into a latticed frond with a personality as delicate as its structure; any slight disturbance will cause this tree to disappear immediately; simultaneous, instantaneous, communal contraction into its base. An unintelligible lump. Here, too, the division of labor, the assignment of castes, the lack of democratic action, as mysterious— more mysterious, even—than the basis on which we decide respective places in our societies. But here, in this colonial protist, we find the first sign of upward mobility. If a tip cell is broken off, a branch cell can take over its architectural and progenitive duties, but only if it hasn't learned too well to be a branch cell. A branch cell cannot remember how to be a leader if it has been merely an ornament too long. Even at this level, we are abysmally ignorant about how cells communicate, but we press on, because such communications are how a multicellular organism is created and maintained.

Cells use many languages to speak to each other; for this there isn't just a single language like the ecumenical genetic code. And we'll have to find more than one Rosetta stone before we can start understanding how cells maintain their near-perfect societies. One bizarre creature who builds a pseudo-organism from many independent cells has been amenable to our probing and prying; at least, it has revealed how it organizes its own cells. That it tells us anything about ourselves is questionable, but just on its own, the actions of these creatures—the slime molds—is fascinating by its improbability and in the eerie unreality of the transformations it undergoes. The slime molds are not particularly slimy (nor probably even molds), chimeras that are variously protists, fungi or primitive plants. Whatever it really is, a slime mold by any other name would be as weird, a tangible living product of a hallucinogenic dream. Slime molds change from amoebas to slugs to mushrooms and back again. Before your very eyes, if you are willing to watch for seven hours, though sometimes they require several weeks to run through their repertoire of roles.

As amoebas, each cell lives on its own, in the soil, creeping from clod to clod, searching for bacteria to eat. But these are greedy, improvident creatures; usually they eat so much and produce new amoebas so rapidly that their food supply has no chance to replenish itself. Hunger. A potent force that can guide any animal's actions. And it is hunger, pure and simple, that turns the slime molds into wandering, many-wiled Ulysseses and begins their odyssey of form and motion. All of the amoebas in the vicinity start moving as rapidly as possible to a central point, a docile mob with none but peaceful intentions. Collecting mysteriously, as gypsies do when their king has died. Although gypsies are said to communicate without physical communication, we have found that the slime molds use more than some form of ESP. When times get hard, one amoeba starts secreting an attractive substance that not only draws other amoebas toward it, but stimulates them to start secreting the same substance from their posterior ends. This substance was originally called acrasin, because, like the cruel witch Acrasia in Spenser's *Faerie Queen,* it attracted creatures (men, in Acrasia's case) and turned them into beasts (a sluggy "beast" in the case of slime molds). Acrasin was at first as mysterious as its powers, then we discovered it was not so very exotic. It is an ordinary, though important molecule manufactured by all cells, "cyclic AMP." Once upon a time this was a very chic molecule; it won a Nobel Prize and was thought to be responsible for almost everything. It isn't. It does act as an intermediate messenger, transmitting messages from certain hormones (not the sex hormones) to cells. These hormones induce the production of cyclic AMP in the cell, and in turn the cyclic AMP activates other enzymes that control the cell's response to the hormone. It also turns amoebas into beasts—a good messenger in both cases. When an amoeba detects acrasin, it moves forward for about 100 seconds toward the source, secreting on its own acrasin that propels other amoebas, ultimately moving them all to the same place, where they gather around a single attractive center.

There is something special about the amoebas that form the center, for if they are taken from the center and placed at the edge, the faithful pilgrims will reverse their former course and move toward their displaced center. Once all of the local amoebas have congregated, a peaceable movable kingdom, they set off on their quest, nothing more than the discovery of what every human being needs, according to Hemingway's poignant story, a clean, well-lighted place.

Slime molds would also like their well-lighted place to be warm and full of bacteria. And for this they will search for weeks, but in the form of a rapidly moving slug. As the group of amoebas changes shape from a congregating blob to a slug with a head, a tail and a purpose, roles of the individual amoebas also change. The original leaders who formed the center of attraction are now dispersed throughout the slug, and new leaders emerge, those forming the seeking head. Who are these? No mystery here; the head of the home-hunting slug is simply the fastest-moving amoebas, for this slug is just a specious, spurious organism. Each cell remains a complete individual, reversibly stuck to other amoebas at head, tail, and sides. The leaders soon find out, as the avant garde often does, that the race is not always to the swift. This is to be their only moment of glory; once they have led their faithful followers to the slime mold version of the Promised Land, they are biblically changed from first to last. They form the supportive but suicidal stalk of the final metamorphosis—the mushroom—in choreographed precision maneuvers that begin when the slug finds a satisfactory place to settle.

The cells in the rear of the slug move underneath the ones at the front which now start to build a rigid stalk. They secrete stiff cellulose walls, making a rigid cylinder that grows upward. As each cell finishes its building task, it dies and falls to the inside of the cylinder that has claimed their lives, unrewarded by reproduction. These must be among the first altruists that ever were. As they build their stalk, other cells from the base enter a transparent globule of mucilage that mounts the lengthening stalk; a shimmering outside elevator to the top. As they are lifted up, these hundreds of thousands of amoebas turn into small dry spores; once at the top, the bubble bursts and the spores are dispersed to new and potentially nourishing environments. When they fall to earth, they change once again into the independent amoebas of yore.

We may never know as much about the development of multicellular organisms as we do about the changes in slime molds. Although spectacular, those changes are immensely simple compared to the prodigious feat of a single unremarkable cell becoming a creature with a brain, an opposable thumb, and three curiously wrought bones against a vibrating diaphragm in the middle ear. To achieve this takes perfect and precise communication, and all of the individual cells must be completely subordinated to the good of the whole; any instructions they receive they must follow implicitly. And there are many instructions. Cells are told when to divide, how many times to divide, what shapes to assume, which directions to grow. A developing embryo of any animal—and especially a mammal—is a mass of complex and convulsive motion, a tiny world in constant upheaval. Kneaded by invisible fingers, the embryonic cells are punched in, rolled over, pushed out, shoved from one place to another. Organs form from a lumpy mush of cells; hearts start to beat, tiny pale stubs grow into arms, then hands and more miraculously, into fingers with fingernails. And, simultaneously, while they are making their morphological excursions, they are changing identities, from anonymous cells to skin cells, liver cells, intestinal cells. But all the time, merely responding to instructions. From the very beginning, while still a simple ball of cells, there are messages; some we can intercept—the electrical telegraphs, for instance. Streams of charged atoms running between adjacent cells.

There are chemical messages, too; part of the brain bulges out, and where it touches the skin an optical lens is formed. Without this touching, there will be no eye. It takes cells touching cells before our spinal cord is formed. We can intercept some of these messages, but we can't translate them. We know what happens when cells touch each other, but we don't know why it happens. In a developing organism, the cells are restless, innately nomadic, and yet extremely polite. If they meet another cell, they stop, refusing to crowd or climb over. To achieve this exquisite courtesy, they have grown a ruffled membrane at their forward end, and this delicate sensor controls the cellular meanderings. If the ruffle touches another cell, its cell is paralyzed motionless until another membrane is grown on another side. It then starts off in that direction, its frilled detector quiveringly sensitive out in front. Perhaps cells come to rest the same way *Paramecia* do. Starting in many directions, stopping only because every direction has been tried, and the conditions there said, "Stop. Go back!" Cells stop out of courtesy for another cell, while *Paramecia* do it purely for their own comfort.

In order to create a new organism, some cells must die, are instructed to die. Only because some cells die are our marvelous flexible fingers and less flexible but still intriguing toes freed from their connecting web of tissue. We walk and manipulate, dance and play the piano, because those cells sacrificed themselves. They are joined in an altruistic brotherhood, perhaps even in a genetic altruistic brotherhood with those cells of *Volvox* and the stalk cells of the slime mold. Those cells, too, are meant to die without a chance to attain genetic immortality; their purpose is simply to build, expendable pieces of architecture in a society whose survival requires just their work, not their offspring. We know that some insect societies are divided up along similar lines; the reproductive elite, sometimes a single queen and males whose only duties are to fertilize her eggs, separate from the workers and warriors who are programmed to preserve, protect, defend, and to die. Mandatory altruism, genetic and unconscious. It doesn't seem to have been absent since the days of *Volvox* and the slime molds. Our entire body is really an altruistic conglomerate, devoted to the maintenance in good condition, of a relatively small elite, the ovaries and testes containing the only cells whose genes have a chance of being passed on to the next generation. Samuel Butler once said that a chicken was merely an egg's way of making another egg, and this is true for every other citizen of our planet, not only for chickens. Kidneys, brains, livers, and hearts and the genes therein are temporal and mortal; only sperm and eggs can taste immortality.

In addition to interpreting all of the directing messages it receives, a cell in an embryo must have, in addition, an absolutely impeccable sense of "self." If we take apart a kidney and a retina from a chicken embryo —really take them apart so that they are no longer organs, or even tissues, but just individual unjoined cells —we can force this strong sense of self to emerge. If the two kinds of cells are mixed together, they neither stay apart, remaining individual cells, nor come together, fusing in an indiscriminate mass of tissue. They move around until they find their own kind; kidney seeks kidney, retina seeks retina. Once they have found each other, the cells start to rebuild the organs. They insist on making their destiny manifest. They were meant to be either part of a kidney or of a retina, and no man is going to put them asunder.

The cellular idea of self is used after birth as well, not to order organs but to protect them. We are so very fragile, really; any mighty subduing master of this earth and all things thereon can be done in by a tiny mindless bacterium or an even smaller, more mindless virus. We are a vulnerable collection of complicated marvels, our organs, products of billions of years of shaping and refining, are vital and irreplaceable. During our time as an embryo, our cells become what they are to become. And that's that. They are once and forever a kidney, a stomach, a muscle, without the easygoing flexibility of the multitalented parts of the Portuguese man-of-war. If a feeding tentacle is lost, that's all right; some other part of that society can take over, happily changing its life-style for one entirely new. Our parts can't do that. Specialization has precluded that, so we have to protect what is our irreplaceable own.

The system that manages that weighs about two pounds and is a collection of blood cells, lymph glands (probably including our misjudged tonsils), and tissues, referred to as our immune system. There is of course no way for this system to tell whether a particular virus or bacterium is going to cause a disease, so it must make the assumption that "anything not ours is bloody," and to be safe, it must destroy anything not ours. To accomplish this, the immune system has to know exactly what is ours and what is not ours; it has to discern self from nonself. And it knows. We don't know how it knows. But it knows, and that is completely unbelievable. Somewhere in us is this phenomenal memory bank where, during the first few weeks after birth, every

125

cell, every molecule, every interior wart and wrinkle
was identified and memorized. A complete, enviable,
permanent knowledge of self. Something, I understand,
that would take us years of abstinence, meditation, and
dedication to achieve. After this record is made, anything nonself that dares set foot in our bodies (virus,
bacterium, protein, sugar, or nucleic acid) automatically triggers the mobilization of our defense system.

The contents of our blood are constantly being
checked by the lymph nodes, intelligence units scattered
through the body. Cells in the lymph nodes compare
the identity of any antigen against their memory of
self; if it doesn't match, some of the detecting cells start
pumping out antibodies at a frantic rate. These proteins move through the blood recognizing and inter-
cepting just the antigen that triggered the system. Their
orders: search and destroy. Encountering their target
antigen, antibodies will bind tightly to it. The binding
itself will immobilize or break open bacterial cells,
which takes care of them. Virus or large invading
molecules are gathered into complicated bundles, to
be collected and disposed of by the trash collectors of
the body, the phagocytes. I am fond of my phagocytes—
really fond of them—those amoeboid cells that wander
through the body, happily cleaning its thoroughfares
and byways by simply eating up any trash they might
find. Damaged tissue, old worn-out, discarded blood
cells, antigen-antibody bundles—these are engulfed by
a phagocyte and disposed of in one hygienic gulp.
Phagocytes are phenomenal; independent, mobile, absolutely essential, and the only sanitation workers who
have never taken advantage of their essential roles to
strike for higher wages.

Nonetheless, admirable creatures and citizens that
they are, I still find their habit of roaming through my

126

body at all hours of night and day a little disconcerting if I stop to think of it. And I do. An infected cut, and I am vividly aware of the phagocytes, alerted through some signal from the damaged tissue, streaming toward that cut, rushing to it as the slime mold amoebas streak toward the scent of acrasin. I realize that they are my own, flesh of my flesh, but they can go anywhere in me they choose. And there are some secrets I want to keep from everyone.

Occasionally, the concept of self operating the immune system breaks down. A tissue or organ is no longer recognized as self and is marked for destruction by the antibodies of its own body. More and more chronic diseases are being explained as such an auto-immune reaction, an anti-self destruction; multiple sclerosis may be one of these. But other than this, the rules of communication and behavior governing the roles of the cells in a body and their interactions with each other are almost always obeyed. Obviously. Whether made up of one cell or many, organisms do not develop into shapeless, disorganized lumps of protoplasm with organs and tissues placed randomly here or there. Mammals almost always are born with four limbs, with five fingers or toes at the end of each, two eyes, two ears, one nose, and so on, and all in a predictable place. Most cells and most creatures have great respect for these rules of communication; their communications straightforward, their submission to the governing dictates unquestioning. Except for cancer cells and man.

Cancer cells do not respect the territorial rights of other cells and refuse to obey the two rules obeyed by all other cells: they neither stop growing nor stop moving when they encounter another cell, and they do not stick to their own kind. Quite simply, they are cells that have decided on autonomy and independent growth,

rather than on cooperation. There would be little in this
to criticize if they were discreet about it. But they are not.
They run amok with as much violence and insensibility
as any Malay caught in that terrifying frenzy. Cancer
will not stop its hideous course of uncontrolled growth
and invasion until it or its victim is dead. Cancer is
illegal and dishonest. It secretes a substance that lures
blood vessels to it; once supplied with its own circula-
tion network, it pirates nutrients from the body, in
greedy and ever-increasing insatiability. It turns in-
vasive, growing into other tissues, dissolving the connections between cells with Samson-like strength. It
can bore holes in muscle and bone. As it divides, its
daughter cells lose more and more of what were once
the fine sensibilities of the cell. They do not stay with
their parental mass; they leave, and totally undismayed
by the fact that they may not belong in a kidney, a
liver, or a lung, they colonize these organs with as little
regard for any of the rights of the inhabitants as the
worst of human imperialists. They grow and grow.
Over cells, and around cells, stealing their food and
space.

I once heard Richard Eberhard read a poem he wrote
after he had seen a tumor, a relative to the cancer that
killed his mother. He was able to find a graphic beauty
in it, and finally surmised that the design of that tumor
would have induced Leonardo to pick up his pen. I
can't think of cancer as beautiful. I think of it as crazy,
pathologically insane. The Turks have a word for it—
delikanli—with crazy blood. They use it to describe
anyone who is exceptionally violent or antisocial. And
the level of violence normal and acceptable to a Turk
is amazingly high. But that's the way I think of cancers—
delikanli—antisocial and utterly, destructively mad.
They cannot listen to what other cells are saying. And

they lie. They manage to convey to the body a completely false message. "I am safe," they say, "I am self." Few other cells can lie. Almost no organism can, except for human beings.

Once upon a time, in the convoluted mass of tissue, fragile and electrical, where we developed our humanness, we also developed speech. Somewhere, somehow, came a fortuitous juxtaposition of physical changes in our larynxes and lips, along with associations of certain cells in the brain. From this combination arose a complex system of communication with other members of our species, the ability to inform them not only of our actions, but of our thoughts. We, above all other animals, have been able to refine our vocalizations from beginning grunts and snarls, to such a high degree of perfection and specialization that our culture and our existence depend on it. We have used it for songs, for poems, to inspire and to comfort. And to lie.

Speech is, of course, not necessary to lie. Other species have signals, some vocal, which they could conceivably use to convey false information to other members of their kind. Because human beings now no longer choose the good of the group over the good of the individual, since we have wrested our futures from the hands of nature who brutally but wisely was able to ensure the survival of the species, we have been able to bring lying to a pinnacle of perfection equal to that of our speech. We cannot *know* whether other animals lie, consciously or unconsciously, but we do know that other species still abide by the main laws of evolution: any innate tendencies damaging the group will be obliterated, and the good (and survival) of the species is the only ultimate good. This is not to say that the ability to convey false information will always be a disadvantage; there are, for example, certain birds who feign injury to entice a predator away from their nests. Thus they protect their young. This is not lying to another member of your own species.

It isn't difficult to imagine that there are tendencies to lie in other species besides our own, but it is difficult to envision a situation where lying would contribute to the good of the species. Among the primates, chimpanzees are well known for their sense of mischief; but in the wild, what would be the result of a chimpanzee who liked to give signals to his troop for leopard or bananas when there were neither leopard nor bananas, just to enjoy the ensuing agitation? Although such a tendency would give a certain degree of immediate satisfaction, especially in a gregarious species pressure would exist for reliability of the signals passed between the members of the group. False signals would develop the potential for ignoring communications from other group members, not only from a prevaricating member. The inability to communicate the actual presence of a leopard or another danger, or of food, would be disastrous to the survival of the group. The needs of the individual, in nature, must be identical with those of the population, and individuals or genetic predispositions which threaten the population are eliminated—in some cases immediately, but always ultimately.

We seem to have escaped, to a large extent, from the harsh laws of nature and of natural selection. It is precisely because we have become concerned with the satisfaction of our own individual needs that we, as animals, can use the technique of the lie with such impunity. In fact, for us, lies have often been essential for survival. In our evolution, the ability to lie and to lie successfully has been a disadvantage neither to the existence of an individual or even to that of a group of individuals. More often than not, survival depends on a lie rather than on the truth. No examples have to be given of the lies that have ensured survival for one group or another of the human species.

Truth has rarely been used successfully to combat mortality. Individual murder, wars, and mass genocide have all been accomplished on the basis of our capacity to tell a lie. Tandem with our inborn capacity to accept authority, this characteristic talent for mendacity that marks our species has had staggering consequences for individuals and races. At a more personal level, the ability to tell a lie or lies, for both the male and female of the species, has many times been direct insurance that their genes would indeed be visited upon the ensuing generation. From Don Giovanni and Faust to King Arthur's sister, duplicity has provided for procreational success. A lesson might be found in the Arthurian legend. Mordred, Arthur's bastard and incestuous son, obviously inherited his mother's talent for deception and thereby brought about the fall of Camelot.

In general, though, our cells don't lie. And we can be very glad of that for the most obvious and practical of reasons; that the bodily society of our cells could not function if there were truth and fiction at the cellular level, and one could not be told from the other. And for the most selfish and abstract of reasons; trying to decipher intercellular intelligences is already too baffling. Knowing that some of the messages we have been able to translate are lies would lead, I fear, to a sudden surge in suicide among cell biologists.

I have an idea that we are neglecting one important area of cell communication. After all, cells are really as distant as stars, so microscopes and telescopes really do the same thing. They both make small objects larger, and show in detailed visibility features invisibile to the naked eye. Optical telescopes were not enough to catch all of the cosmic colloquy; it was only when we started aiming radiotelescopes extraterrestrially, magnifying the sounds of the universe, only when we started to listen, did we make real progress. Listening, we could hear the pulsars, quasars, and black holes out there. Cells might also communicate to each other by talking; coded speech of sound as well as of touch or chemistry. What we need is a radiomicroscope, so we can eavesdrop on the tiny, vital gossip of life.

SEVEN

Energy

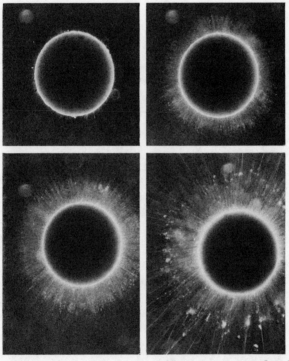

(courtesy of Prof. Karl Grell)

Plate VIII. *Thalassicolla nucleata,* a "sun animal" (heliozoan). The corona of projecting rays is actually composed of extensions of living protoplasm.

I clearly remember being introduced to the neutrino. I loved the neutrino, though not to excess, of course. No mass, no charge, I was told, just pure energy, the end of a long road of energy conversions; and it could not be used again to create matter, this neutrino— at least in the universe we inhabit. Perhaps somewhere else, in that postulated universe, next door but invisible, the one made of antimatter, neutrinos could be used again to build matter, but not here. Neutrinos the ultimate nihilists. I was fascinated, especially by the so-called "neutrino sink." I pictured it as a large hole, somewhere out there with the exotic elite, the neutron stars and pulsars, the quasars and the Crab Nebula. And masses of neutrinos spiraled down into it in the quintessential, final vortex; the energy in our universe draining away forever. I had hoped that this wasn't completely true, I wanted my neutrinos to be recycled, even if they were tossed on some interuniversal compost heap to be used in a time and place I could never know. That view of the neutrino, in fact, wasn't completely true; I wanted my neutrinos to be matter and merge with it, but it is an improbable and rare event. Only one out of every ten billion neutrinos that pass through the earth ever collide with one of its atoms. I hope the rest collide somewhere in space or in another universe. My atoms and their subatomic particles have been a long time coming to me, and to think that they, and their memories (especially one particular bottle of Haut-Brion, 1953 I think it was) will be wasted, just lying around all inoperative energy, a good deal of past but no future is rather depressing.

I don't believe that is the way it happens. First, I am sure that we don't know everything; messages from outer space keep telling us that every day, bringing us information that we did not predict and cannot explain. Second, this universe—at least the part of it that we are part of—is a saving, parsimonious place; the original "use it up, wear it out, make do" pattern for the frugal New Englander. I can't be authoritative about the universe, but I do know about life on this planet, and there is every reason to believe that it follows a universal generalized principle—or set of principles— in its actions. If so, it follows a saving principle, for the system we are part of is a constantly recycling, re-using, stingy one. Plants, animals and everything else alive are fed, clothed, and heated by the sun. All of this couldn't be managed without the greatest economy and exchange. The energy that our cells use to keep themselves in order until chaos finally prevails, comes from plants who get it from the sun. Wool and silk are made by creatures that feed on plants, and use plant proteins to build themselves shelter and protection from the elements. Cotton, linen, nylon, rayon and dacron: all are fibers from plants; the last three are made from the sunlight that shone in the Paleozoic, by plants that did not look like today's plants and were not, but were their ancestors.

The Egyptians did not invent nylon or gasoline, but they were clever; they managed birth control and boomerangs. They also came up with the central idea of the energy cycle we are in, they acknowledged the sun as the Giver of Life. We don't call it Ra, we call it photons; and without it we would not exist in any form. Now, the sun does not give us all of our atoms; the ones heavier than iron had to come from larger stars somewhere else. In fact, the sun may have given us none of our atoms, but it gives us all we need to furnish those atoms with order, to put them together into molecules. Whatever we might have gotten from other stars would be useless stuff without the intervention of our star, the sun. We need the sun so we can hammer our molecules together and give these molecules the energy to stay together once they have been formed into organisms. The first great step—the creation of this universe—is incomprehensible; the first great step in our own little microcosm, just the sun and its planet earth, is more accessible to understanding but equally impressive.

The transubstantiation of light into chemical energy —energy that will become food for everyone—is not of epic dimensions, but spectacular on its own. In the early years of life, the discovery of this secret alchemical process (true alchemy this, transforming something common into something precious) made cells independent—freed from the vagaries of random nutrition, life and growth dependent upon whether or not something useful drifted by in their liquid homes. Plants have inherited the secret and the independence. To feed the entire planet, every one of them, from microorganism on up the ladder to human beings, all plants need is a short list of sixteen everyday items: carbon dioxide, water, light, phosphorus, nitrogen, potassium, calcium, magnesium, boron, sulfur, manganese, iron, zinc, copper, molybdenum, and chlorine. With just this brief inventory and using less than one percent of the sunlight that hits the earth, plants supply the needs of everyone; a modern re-creation of the multiplication of loaves and fishes. Some are supplied less well than others, admittedly, but that is the fault of politics, not the fault of photosynthesis, the biochemistry whereby our sunshine is turned into our food.

There are about seventy separate chemical reactions involved in photosynthesis. Those seventy reactions, besides feeding and clothing us, warm us and move us

137

from place to place. And the extraordinary combination of them evolved through chance, in combination with atomic consciousness, perhaps. And with time. As George Wald has said, "Given so much time, the 'impossible' becomes possible, the possible probable and the probable virtually certain." And it may even be that what we think is impossible is not only possible but highly probable, if our atoms have some idea of where they came from and what they are to do in the future. Whether that idea is actually true or merely probable, photosynthesis, probable or not, is truly wonderful. Actually, every step after the first giant step—the actual transformation of the sunlight—is straightforward biochemistry and not so amazing. But that first step. Even knowing almost completely how sunlight is changed into chemical energy does not lessen our conviction that it is truly a miraculous event. The photosynthetic molecules—chlorophyll, for the most part—manage this transformation. We paint hospital walls green and clothe surgeons and nurses in the same color, I am told, because the human brain finds green relaxing. I wonder if green is so comforting because of a long-ago memory we keep, that trees mean safety, a memory made when we were arboreal tailed creatures, still quadrupedal but with a great future before us. Or perhaps green makes us feel calm and safe from an even longer-ago memory, from the glistening green of chlorophyll, the reassurance of food, of plenty, always there, except during the winter. The green chlorophyll-containing chloroplasts signaling the presence of the vital links in the infinite earthly cycle of survival.

We play our part in this cycle, just as the ancient Greeks thought, synthesized from fire, earth, air, and water. The fire of the sun, carbon dioxide and nitrogen from the air, water, and the earth: inorganic materials, products of rocks rather than of living things, but taken from the soil by the plant. And all of this—fire, earth, air, and water—form a constant cycle of synthesis, breakdown and loss, then resynthesis. The solar energy we feed upon is ultimately nothing more than heat sent back out into space to be used for unknown purposes. Sometimes the energy may be trapped for only a short while, in the bodies of plants; for longer, perhaps, in the bodies of animals; but always liberated after a while by death and decay. But it can also be held for millions of years before it is released. It has been a long time since the Paleozoic, when, during the sixty-five million warm moist years of the Carboniferous, there was worldwide massive photosynthesis and energy storage. The plants were large, exotic, and so abundant that certain coal deposits are made just of the spores of these plants. And as plants left the monumental stores of their carbon products, small-brained, four-legged creatures wambled through these alien forests. The carbon was compressed, and the spraddle-legged creature changed and changed again. One recent change brought, after a few hundred million years, a two-legged creature who in little more than fifty years has brought to a visible end those once-vast deposits of energy. In about fifty more years, that puny David will have completely pirated those millions of years of patient work and storage of carbon-held energy in coal, gas, and that dark greenish-brown foul-smelling liquid he is sure he cannot live without: petroleum.

It all began with photons. The fastest things in the world, photons, particles of light traveling, of course, at the speed of light. They may also be among the lightest things in the world. They may not weigh zero, but they can't weigh much more than that. They carry an immense amount of energy for their nearly nonexistent size because they are speeding about so. They run into things, and the collisions we benefit from most are those with the verdant molecule, chlorophyll. There are other light-absorbing pigments in green plants, and the photons can hit these, but instantaneously this energy is transferred to chlorophyll, the key molecule in green plant energetics. It is energy we are after here, and also a way of keeping that energy from vanishing before the cell can use it in its constant fight against encroaching chaos.

The best way to keep energy stable and available is to lock it between two atoms, as they bind together to make a molecule. Atoms are yoked together through a sharing of each other's electrons. Instead of just sprinting around the nucleus of their own atom, the electrons of two joined atoms include the nucleus of the other atom in their orbit as well. This spins a glue of energy that keeps the two atoms together. In some cases, energy is released when atoms are bound together, but in others energy must be added. The energy needed to bind the atoms and keep them bound is held by the atoms until the bond is broken, imprisoned in the orbits of these racing, spinning, mutual electrons.

One of the many chemical chores of green plants is to put energy between the carbon atoms that make carbohydrates; that is photosynthesis. It takes just air, fire, and water for photosynthesis; the plants have to go underground for everything else other than carbohydrate synthesis. The earth provides the arena for the remaining food-gathering activities of the plant, the absorption of atoms from the soil through the roots.

The most important energy-holding molecules, although other carbohydrates are built with solar energy, are sugars—especially glucose. Before a cell can release energy from other carbohydrates, it must first change them into glucose. Breaking down glucose to get at its energy-rich molecular marrow is the function of respiration. It takes our cells—and those of the plant—more than eight separate reactions to release all of the energy held in a glucose molecule. Plants put the solar energy into glucose through photosynthesis, but they also respire as we do, breaking down the glucose molecule with exactly the same complex chemistry, breathing in oxygen and breathing out carbon dioxide. For each human being, plants have to fix about 260 pounds of carbon into carbohydrates each year to allow us to support our 60,000 billion dependents, the cells that manage our lives. The energy in the molecular bonds between all of those carbon atoms is only a part of the grocery list; our cells also ask for proteins, fats, and nucleic acids, and the plants supply those also.

The energy is ensnared within the chloroplasts, bright beautifully green rounded structures, stacks of membranes held inside membranes, whose worth is far above the emeralds they copy. Once these chloroplasts were independent blue-green algae, and they still manage independent movement, even within the confines of their directing cell. Plants and their leaves will orient themselves so that the upper surface of the leaf is collecting the maximum amount of light, twisting and shifting to put their solar collectors at the best possible angle to the fall of the sunlight. Inside the cell, too, there is such sun-seeking motion, compensatory motion for the motionless containing organism, held in one place, tied by the hundreds of mouths that reach through the underground, the roots that pull in water and minerals the air cannot provide. The green chloroplasts can twist and turn on their own, gyrating inside the cell until they find themselves fully facing the sun, catching as many photons as they can for their chlorophyll.

Chlorophyll is the prime member of a molecular elite; those molecules that can actually do something with the sun. Many molecules can reflect it or absorb it; the dark brown melanin that colors our skin and eyes, the hemoglobin that incarnadines our blood, and absorbs all but the reddest wavelengths of light, but can do nothing more with it. The carotenes that color squash yellow and carrots orange absorb all but the orange and yellow wavelengths; but common as they are, they share the uncommon capacity of chlorophyll to do something constructive with the light that brightens them. This talent lies in the structure of these complicated molecules, for one part is made of carbon atoms joined together with a special kind of bond; in these bonds the electrons can be shared ultimately by all of the carbon atoms in the particular substructure. The electrons vibrate back and forth, resonating between one carbon atom and another, associated yet free, restless and nomadic. These electrons are more free than almost any other electrons; they are free to stay with their molecule or free to leave it completely. If one of these special molecules, a carotene or a chlorophyll absorbs a photon of light, the energy can be transferred to one of the electrons in the resonating coalition of carbon atoms. This extra injection of energy excites the electron and allows it to break free of its original molecule; it can take up with another molecule and bring this extra energy with it.

In photosynthesis, solar energy is converted and used in this way. The energy from the photons is given to electrons, which are in turn passed to one molecule after another, transporting that energy with them, until they finally leave the energy in a molecule that can participate in general cellular metabolic reactions. Two photons of light are required to move each electron this way, and also a very real amount of energy— 1.2 volts—but that energy can be stored finally between two carbon atoms in a carbohydrate. In green plants, the chlorophyll is helped by other pigments—carotenes and carotenols—that capture the light but funnel the energy to the chlorophyll so it can excite its electrons. Only a small proportion of all of the chlorophyll molecules in a plant send their excited electrons on to other molecules, about one in every three hundred. The others are antennae, they collect the light, but funnel the energy to this one chlorophyll liaison with the general cellular metabolism.

Electrons can hold onto the extra energy they have been given for only a very short time; the excitement of it makes them unstable, and they will very quickly release that energy. It is a long road, chemically, from excitation, when one of the photons gets trapped by the chlorophyll, to where the energy can be held for a while in a carbohydrate's carbons. Even the fast-moving electrons are apparently unused to excitement, for they can only hold on to the energy for 10^{-15} seconds, not even a fraction of a second, really, just one million-billionth of one. In that almost immeasurably short time, light itself can travel (in a vacuum) only 3/10,000 of a millimeter and it is the fastest traveler of all. Yet the electrons from chlorophyll are tenacious compared to those in carotene or carotenol; their electrons lose their energy so quickly we cannot even measure it. As short as that million-billionth of a second is, incomprehensibly short, the excited electron can still move through a series of molecular stations. The first

143

stop was of course in a chlorophyll, one identified by the number 680, because it primarily absorbs light of that wavelength, reddish-orange (680×10^{-7} cm). The electron is excited with the energy from the two photons and sent on to a molecule, so far identified only as "Q." As chlorophyll loses each electron, it is replaced by an electron taken from another incognito molecule, "Z," that gets electrons from hydrogen atoms that have been part of a water molecule. (The molecular interactions of photosynthesis are somewhat complicated, but far easier to follow, I am sure than the circumlocutions and incongruities of a daily soap opera.)

Hydrogen is a simple atom, the simplest. Around its nucleus of a single positively charged proton orbits a single negatively charged electron. The mysterious "Z" separates the electron from its proton, and only the electron—because of its energy-carrying capacity—is sent on to chlorophyll. After the electron has been transferred from chlorophyll to "Q," like a subatomic bucket, it is passed on again to and through an electron carrier system composed of several more molecules. It then begins a second stage of its journey with a very brief stop at another chlorophyll, whose number is 700. Chlorophyll 700 also is organized with hundreds of antenna chlorophylls funneling solar energy to the one reactive chlorophyll molecule. So here the traveling electron gets an energetic boost before it continues on its way to the third anonymous molecule of photosynthesis "X." After its brief encounter with "X," the electron is reunited with a proton to form a hydrogen atom, and as hydrogen carries its precious burden of solar golden energy into (at last) a stable molecule that will hold the energy until it is needed by the cell. The entire transport of the electron is completed in that less than a moment of excitement, before the electron

144

can lose the energy it was originally handed in chlorophyll 680.

This energy will be used primarily to bind hydrogen atoms to molecules of carbon dioxide and to bind these carbon atoms to one another in a carbohydrate. In glucose there are six carbon atoms, and six bonds riveting those carbons in place. An amazing amount of energy is stored in that one sugar molecule. In each gram of glucose, there are 3,800 dietary calories, more than enough for the energy needs of the cellular population of the average human being. The plant uses some of the carbohydrates it makes for its own survival, breaking them down, as we do, through cellular respiration, into the component carbon dioxides and hydrogens.

But plants make far more carbohydrates than they will need, and the extra trapped energy is kept in long chains of glucose molecules, either as starch or as cellulose. Every year about 150 billion tons of carbon are fixed by the green plants of the world into molecules like these, that will supply the energy needs of the cells of all living things. The use of this energy, the passage through a food chain consisting of plants and animals, has been called terribly inefficient by efficiency experts. Look at the terrible waste, they say; it takes 100,000 pounds of photosynthesizing marine algae to be converted into just one pound of codfish. But along the way that energy has nourished smaller animals than codfish, although a tremendous amount has been lost as heat. But does it matter whether it's an efficient transfer? It's all free food, you know, except for those animals that have been so unwise as to set up a money-based economy. To say it is inefficient is to say nothing, for we are able to use only a tiny amount of the energy hitting the earth, and there is no way we can persuade the sun to burn more brightly or less brightly. Anyway, the transfer of biological energy is just as efficient as our pride, the internal combustion engine, and far more efficient than the incandescent light bulb. Human beings should tend their own energy gardens; plants do as well and are willing to distribute their products without charge to the public.

Getting the energy out of glucose takes about as many steps as it took to put it in. For a cell, 3,800 calories (actually 3.8 heat calories) is a tremendous amount of heat; too tremendous. If freed all at once, the cell could not use it; it would be dead. The enzymes would be cooked into inaction, and not a single biochemical life reaction could occur. Energy release has to be done step by step, dismantling the glucose molecule one carbon at a time, for the cell can handle that amount of energy. If the energy were just freed indiscriminately, it would be totally useless to the cell. It has to be stored so that it is dispensed only when needed, and transported to the places within the cell that use it. Only a limited number of molecules can manage this, keeping energy stable but available, and they can do it because of special molecular bonds, high-energy bonds easily broken. There are just a few denominations of this energy currency, universal and cellular. The one most used is ATP (adenosine triphosphate) and it was a fortunate day when cells happened upon this particular molecule. Actually, this molecule had to be discovered before there were even cells, when there were just systems, for there is no cell that transfers energy without using ATP. Like the nucleic acids, it must have been one of the earliest successful molecular experiments made. Any time a cell needs energy, ATP is there; as muscles contract, as electrical messages are transmitted along the delicate axons of a nerve cell, when molecules are pulled into or pushed out of a cell.

ENERGY

We measure everything we do in terms of dollars, and cell activities are measured by the number of ATP molecules the activity requires or yields, how many ATP molecules can be built from the energy released in a reaction. Before there was oxygen in the world, only very little of the total energy in a glucose molecule could be gotten out and used by the cell. The chemical reactions that were used, the best these cells could manage then, still form the initial steps of our respiration, inefficient heirlooms perhaps retained for sentiment's sake, they aren't kept because they are practical. These first few steps rearrange the glucose molecule and then split it in half; the energy is then in packages of three carbon atoms linked to each other rather than six, a molecule called pyruvate. All of these steps have to take place in the absence of oxygen, and were all the cells had before the Prometheus mitochondria brought to the cell the power of completely burning the glucose by using oxygen. These ancient steps could lead to either of two destinations when there was no oxygen—alcoholic fermentation or the formation of lactic acid. In fermentation, the pyruvate loses a carbon dioxide, and is changed to alcohol; Bacchus evidently came before Prometheus—we warmed ourselves with spirits before we warmed them by burning. This particular pathway is still used only by the yeasts and by a few other microorganisms.

The second pathway, where pyruvate is changed into lactic acid, is still embedded in our cell chemistry. If we move about much at all, we are familiar with these biochemical steps. Painfully familiar. If we move our muscles faster than our heart and blood can supply them with the oxygen they need, pyruvate becomes lactic acid, and lactic acid in our muscles is the feeling of tiredness and fatigue. This is oxygen debt; our mus-

147

cles will still work, but there is not enough oxygen to allow pyruvate to continue through the complete respiratory cycle. Pyruvate picks up hydrogen, becomes lactic acid and we ache.

Both fermentation and the conversion of pyruvate to lactic acid are inefficient, providing the cell with very little energy in the form of ATP. To those cells whose ancestors once gave refuge to the microorganism that later became the mitochondrion their guests gave a gift every bit as important as the one from the original Prometheus, and the givers were punished far less. They lost only their freedom, and were not unkindly chained to a stone to have their livers plucked from them by an eagle, bit by grisly bit. Pyruvate is taken into the mitochondria, and only there can the final stages of the complete burning of glucose take place. The pyruvate is sent through a long series of changes and rearrangements: it is combined with another molecule and then broken down, carbon by carbon. As each carbon is broken off, carbon dioxide is released, back into the original form in which it was used by the plant.

Energy is liberated as these bonds are broken, and placed into ATP. At certain places, the hydrogen is also broken off the molecule, and it is sent around from molecule to molecule, an Alice in Wonderland reversal of the photosynthetic transport that hydrogen atoms undergo during photosynthesis. This respiratory hydrogen is broken, as it was by the incognito "Z," during photosynthesis into an electron and a proton. The electron is sent through a chain of molecules, transferred from one to another. This is the respiratory chain, a life-saving chain for cells that have to live in the presence of oxygen. Through the mediation of the molecules in this chain, oxygen combines with hydrogen to form water, rather than forming hydrogen peroxide, deadly poison for almost any cell.

The enzymes required for the transport of the electron through the respiratory chain all seem to be located together on little knobs projecting from the inside of the mitochondrion, conveniently organized for the faster than rapid transfer of the electrons that has to take place. In photosynthesis, as the electron is passed around the various molecules, energy from the sun is used to push it through; it is an uphill journey, requiring energy. However, when the electron is passed through the respiratory chain, it is all downhill. At each step, the electron gives up some energy, a bucket spilling some of its contents every time it is moved. With oxygen around, about 65 percent of the energy held in the glucose can be freed for use by the cell; the rest is lost as heat. And, at the end of this effective respiratory chain, the electron is again reunited with a proton, to become hydrogen.

Plants and animals break down their carbohydrates the same way. We can't use the carbon dioxide that is made each time a carbon atom is broken off, but the plants can; they just shove it right back into their machinery, but into the photosynthetic part. A carbon's work is never done. Just as soon as it has completed its job, carrying the energy from the sun to us, before it even has a chance to rest, it is pulled into a plant cell, to be recycled into glucose or some other carbohydrate.

The same is true for hydrogen atoms. After the electron has yielded all the energy it can, carefully parceling it out in usable quantities, its job is still not done. Before it even has time to enjoy its reunion with the proton, it is seized by the oxygen we have breathed in through our lungs and have brought to our cells by our blood. Oxygen is a molecular mop, cleaning up the hydrogens released at the end of the respiratory chain, combining with them into water. That is respiration—all of this—why we breathe in oxygen, breathe out carbon dioxide, and eat.

To live, most cells must use this respiratory chain; cyanide poisoning, quick and deadly, illustrates what this chain means to us. Cyanide looks like an electron to a cell (after all, none of us is perfect, even though cells come closest). Their cellular myopia is fatal. The cyanide takes the place of an electron in the respiratory chain, but it won't be handed on. Respiration is blocked. The organism dies. For a constant cycle of energy exchange must take place in the cellular world as it must in the entire biological world: fire, earth, water, and air in their eternal flux.

Although plants supply us with all we need, most of their carbon is fixed into cellulose, and we cannot get at that energy directly. We cannot get it out by merely eating the plant, for like every other animal, we don't have the enzymes that break the carbon atoms apart the way they are held together in cellulose. Only a few bacteria have those enzymes. We own some of these bacteria, down where we hardly notice them, in our lowest intestine, the colon; they do break cellulose down and make its carbons available to us—it's just that it happens where the talents of the bacteria can do least possible good. There isn't enough time for us to take advantage of them. There are some animals—cattle, sheep, goats, and deer, among them—that have evolved a special internal agriculture that makes the energy in cellulose available to them. At the beginning of their digestive tracts, they have a large, useful chamber, the rumen, populated by masses of these cellulose-digesting bacteria. The bacteria are happy; the animals are providing them with large supplies of the cellulose they like. The cattle are happy; their contentment coming from the knowledge that the cellulose solar energy is being freed for them and that they can get the proteins and nucleic acids they need from digesting the rumen bacteria themselves. Before we even had the idea of domestic agriculture, cattle and sheep developed grazing herds of their own, intestinal cattle that can be used as we use a grazing steer, to digest foods inedible or unpalatable, and convert it into useful food.

Man does not live by carbohydrates alone. We need protein. This is perhaps the most cooperative food venture, for it not only requires cooperation between the air and the earth, but also another cooperative effort far more strange. Making proteins, or even their subunits, the amino acids requires nitrogen. Nitrogen is plentiful, but only as gaseous nitrogen in the air, and plants cannot use nitrogen gas, any more than we can. They need to have it combined in either of two forms— nitrate or nitrite—and it has to be in the soil where they can pick it up through their roots. Lightning can take nitrogen from the air, and plant it in the soil as nitrate and nitrite, but lightning is hardly a universally constant or predictable fact of life. It takes pleasure from its inconstancy and capriciousness, we are told. So again it falls to the plants to solve the problem and feed the world its proteins as well as its carbohydrates. And plants have solved it in the most unlikely way. But really unlikely only if we hadn't learned the symbiotic secrets of the protists, how prokaryotes moved into the cells bringing cell division mechanisms, the respiratory chain, and photosynthesis. A common protist solution to a lack of vital talent: just form a permanent association with a smaller organism that can provide what you lack.

THE CENTER OF LIFE

A few groups of plants have discovered the joys of symbiosis, much to their delight and to ours: the tropical grasses, which are not important to us; and the legumes (peas, beans, clover, alfalfa, lupines), which are important. Virtually every protein or amino acid we eat comes from legumes originally, for they alone can take gaseous nitrogen from the air and fix it into nitrates and nitrites. But they cannot do it alone. Certain free-living bacteria, whenever they have the opportunity, will take up residence inside of the roots of legumes, and once there they take nitrogen and change it into forms that the plant can use to make the amino acids it needs for the manufacture of proteins. The bacteria are the partners that have the genes specifying these nitrogen-fixing enzymes, but the genes are not expressed until the bacteria are tucked into the oxygen-free environment of the roots of the legumes. We either eat legumes or they die and decay, so their nitrogen compounds merge with the earth, to be there for the use of other plants that cannot get these compounds on their own. Thus are we provided with the proteins we need, completely dependent for that part of the diet on plants, and just on that one group of plants—the legumes with their invaluable immigrant bacteria.

After *Mixotricha, Paramecium,* and the other protists that did so well by adopting symbiosis, the legumes' solution is not a particularly surprising development, even though in this case it was an entire organism that formed an alliance with a prokaryote. It was an impressive solution; before the invention of the Haber process, which makes nitrates and nitrites expensively from natural gas, these symbiotic associations were our only nitrogen fertilizers. They cost us nothing and we did all right.

What is completely surprising is the hemoglobin. When you pull a legume from the soil, there it is— hemoglobin staining the earth as red as it stains our blood. Almost identical to our very own hemoglobin, even though that molecule is inscrutably and inexplicably missing from any plant or animal evolutionarily intermediate between pea or bean and vertebrate.

Nor was there hemoglobin in the bacteria who are even further away evolutionarily. They put it together cooperatively and exclusively (outside the vertebrates), the legumes and their bacteria. The plants code for one part of the molecule in their genes; the bacteria have the information necessary to make the other part. They have been forced to make hemoglobin because they need it, not to bring oxygen to their tissues as it does so faithfully and well to ours, to mop up the wastes of respiration, but to keep oxygen away from the plant tissues. Hemoglobin is hemoglobin, and this legume molecule picks up oxygen as well as ours does, but it doesn't release it; it holds on to it, to keep it away from the nitrogen-fixing enzymes that are inactivated by free oxygen.

So this eternal cycle requires hemoglobin, and in the most surprising places. It also requires death and life. Plant life and death. And our life and death. Plants must be alive to photosynthesize, and to remain alive, animals must eat the plants or each other. But everyone involved has to die and decay, to return to the earth and air the atoms that new plants will need. Everyone must be finally disassembled by fungi and bacteria into component atoms and molecules. The solar energy that has put these together, whatever remains of it, is radiated as heat out into space, leaving this earth for another—perhaps ending up in neutrinos that are reassembled in some other world.

THE CENTER OF LIFE

One part of this universal scheme we like. The other part we don't like at all. We like the part when we are alive, when we eat and think up new ways to use the plants' contribution, when we manage to sequester enough of their atoms and energy to make ourselves comfortable and to move around at our appointed tasks. We like to borrow the atoms from the various parts of the energy cycle. We like this part so much we are determined to turn as much of this infinite cycle of fire, earth, air, and water to the profit and use of a single species. Us. We really don't mind borrowing, but we are certainly reluctant to pay the loan back. We refuse to decay naturally and return our atoms—useless after our death until they have been replaced in the cycle. Partly due to the morbid nudgings of our undertakers, we lock our bodies into metal coffins, wrap those in concrete vaults, and defy anything natural to happen to our lifeless bodies. Fungi, bacteria, and the other decomposers are to be kept at bay. When I think that in every acre of soil we use, there are three tons of these little fellows doing their best to support our farming, I think we are immensely selfish.

EIGHT

Death

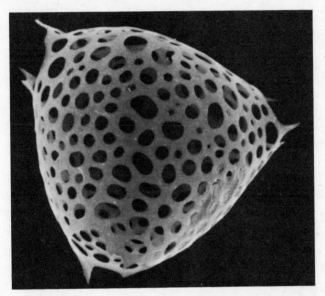

(courtesy of J. and M. Cachon)

Plate IX. Universal geometry—a tetrahedron, designed
and built by *G. Collosphaera*. Magnified 600 times.

155

W e all have to do it—die. Young and old, those whom the gods love and those allowed to fulfill their full appointed time, we all come finally to our own *dies ira, dies illa*. Day of judgment, day of wrath. And we don't like it, most of us. Plastic surgery, wigs, toupees and makeup help us enter the make-believe where we pretend not to age; and we lock our personal *memento mori*, an aged or dying parent, away where even scarce visits seem too much for us to bear. It's not something we want to be reminded of, *ever*. Our steady journey away from youth and toward mortality, a journey that always for everyone has the same end. We can't escape; our bodies are programmed to run fatally down at just about the biblical allotment of three score years and ten. And not all of our piety and wit, all of our great, superb advances in medicine and technology, have been able to grant us much more than that. The human life span alloted by our genes has not increased since biblical times; we all start aging about the same time, and most of us are dead by the time we reach the seventy-year mark. Medicine has just given more of us a chance to hit that seventy-year mark. Medicine has added, they say, twenty years to our lives, by the advances in treatment: the defeat of infectious diseases and the discovery of anesthetics and surgical techniques. But that twenty years just carries us closer to—and not past—that solid, impenetrable threescore years and ten. Some people do live to be one hundred years old, or even a few years past that, but that was true in the time of Julius Caesar. It was just that then there was more chance of being ripped untimely from our earthly womb by plague, smallpox, malaria, or an assassin's dagger.

This famed twenty years is very like a Trojan horse: attractive only from a distance. We see that it is really just twenty years more to fall prey to the thousand natural shocks the flesh is heir to. We don't die much anymore from pneumonia and diphtheria, that's true; but we still die, and we still die at around seventy years of age. We don't get twenty more years of youth from medical science, or even of middle age. We get twenty more years of the unenviable effects of aging on the human body. We get more time to suffer from as yet undefeated heart disease and cancer, today's major causes of death. And even if they were defeated, we would still age, our joints would get stiff, our senses dim, our energy fade. The inescapable, unfaceable fact. Our bodies will age and die long before we want them to.

Why? In evolutionary terms, it is clear why we die. We die because we are not supposed to live. We are useless to the species after we have reproduced and made our genetic contribution. The only talent honored in nature (which is devoid of the finer sensibilities, among them respect for age or personage) is the ability to reproduce. Nonreproducing organisms are of no use whatever to evolution and natural selection; they are just around, clotting up the landscape and perhaps taking food and space away from some younger, upcoming and more successful genetic experiment. Genetic experiments that have to mature first, and that can be terminated and their genes lost as easily as male Serengeti lions, new to the pride, break the necks of the cubs fathered by their predecessors. One strong, swift shake, and the genes of the males they are replacing are gone for good from the pride, making room for those of the new males.

We have never had to remain at the peak of our reproductive capacities much past the first thirty years of life, during the several million years of our evolution, simply because we rarely lived even that long. There was no need for healthy old human beings; any adaptation to that end would have been wasted genetic effort, for there weren't any old human beings. Long before we got old our skulls were cracked by another human being wielding a thighbone, or we were eaten by a cave bear or trod upon by a woolly mammoth. For whatever reason, our lives ended in those days; our genes never got a chance to happen upon a solution to aging, and pushing them away from it if they should have happened to get too close to one, was the fact that old animals are useless to the species. Completely expendable. We know that animals that reach a certain size start aging when they stop growing at maturity; that is programmed in our genes too and before now we never got a chance to get rid of that particular set of instructions.

It never before has mattered that our cells last us much past thirty. Up until then they do a fantastic job. Human males are at their prime of life between the ages of nineteen and twenty-five. They are faster, stronger, more reproductive, and more interested in reproduction than they will ever be. In a very short time after that, we reach the age of thirty. At that point it becomes more probable that we will die than that we will continue to live. That unhappy probability just keeps increasing, doubling every eight years we live after thirty, until it finally becomes highly improbable that we will live longer. Finally, we don't. These are odds that no one has ever been known to beat. To some, this general deterioration in our capacities as well-oiled fighting and loving machines is evidence not of neglect on the part of natural selection, but evidence that we are purposely programmed to die. We are meant to die; a planned, plotted obsolescence and removal to allow room, more room for the following generations.

However we look at the actual fact of death, as an accident or a specific intentional genetic directive, we still see that we must die. For those of us who would rather stay alive, we seem to be cherishing an unnatural desire if we accept our role as one of a million species mutually connected and subject to the same laws of natural selection. And more than this, we not only do not want to die, we want to stay arrested at a certain age; we don't want to live as an old person, we want to remain a young healthy sexual animal.

To fulfill these desires, we must renounce our evolutionary connections with the living world, forswear ourselves as a natural species, evolving according to the laws followed by every other living thing. We seem willing to do that, unaware that to be both immortal and sexual are contradictory impulses. For not only are sex and death closely entwined—love and death were inseparable long before any sensate novelist thought fit to point it out—but with sex there also has to be death. In the natural order of things, amoebas do not die. They may get eaten or heated to oblivion in an evaporating puddle, but they do not age nor do they die; this is because amoebas do not have sex, and so each amoeba has a chance of achieving a kind of cellular immortality. With time, an amoeba does not get old; rather it rejuvenates itself, but the requirement for drinking from the fountain of youth is the renunciation of sex. Amoebas grow to a certain size and then they divide into two new amoebas, as newborn and as young as an amoeba can ever be. These new amoebas, are identical in form and genes to their single parental amoeba. Thus the parent amoeba no longer exists. But it did not die; it has been born again through this cellular fission. Since there are amoebas—many billions of them today—there must be somewhere in this world at least one amoeba that has been alive for several billion years, showing no effects of its long, interminable life.

With sex there must be death for there to be the continuing life of the species. To make room. Room for the products of the incessant genetic experimentation; the new forms that have arisen from the mixing and remixing of the genes that sex achieves. *Paramecia* have advanced far beyond the asexual immortality of amoebas; they are completely adjusted to and totally dependent upon sex for their survival. *Paramecia* can and constantly do, reproduce asexually through cell division. But, in the absence of sex, a *Paramecium*—or rather a *Paramecium* population, grows old. The rate of reproduction (division) decreases, the death rate increases, the *Paramecia* become thin, inactive, pale, and wan. *Paramecium* stocks are usually maintained as unisexual populations. To keep them this way, asexual and isolated in tubes or flasks, requires periodic ritual sexual matings to rejuvenate the stocks; otherwise they will languish and die. It could be from loneliness, that is true, but more likely it is because they need a regular renovation of their genes, which only sex can give them.

Perhaps it is not only *Paramecium* that can need the rejuvenating capacities of sex. I have a sad memory from one summer symposium at Cold Spring Harbor. There among the usual euphoric stimulation of these annual symposia, I first realized the cheerless, sterile fate of many a woman scientist. I roomed for one night with another female biologist, a few years older but similarly unmarried. Apropos of nothing, this woman began to bewail her single state, and with the saddest mirth I have ever heard, she quoted one of her colleagues who had assured her, quite cruelly that women scientists were like the cells of a multicellular organism; once they had specialized, they did not reproduce. Specialization has indeed caused aging, a strange sterility and a pathetic sort of social death in many (not all) of the women scientists I have known. The biological fact behind this unkind comparison may even be a partial explanation for the aging of us all.

Cell division or cell reproduction allows cells to keep young, as the amoeba does, and also is necessary for the repair and replacement of tissues, as life takes its toll in general wear and tear. But the cells that make up most of our organs, except for the pancreas and the liver, fall into the same category as those specialized female scientists. They do not reproduce; once they have achieved the highly specialized state necessary to fulfill their intricate tasks, most cells in our organs do not divide. They cannot even replace their tissue if it is damaged.

Injured stomachs, intestines, kidneys, or hearts cannot replace the damaged cells except by nonfunctional scar tissue. This reduces the overall capacity of the organ and places a larger burden on those already burdened healthy cells that remain. It may mean that the remaining cells age more quickly. This is the price we must pay for those wonderful organs; all of their component cells have their genetic attention focused on doing their jobs, there is none left over for reproduction. Our marvels of evolution, our intricate organic machines, unsurpassed by man-made machines, are basically just as sexless and genetically hopeless as a worker bee or a soldier ant.

Not our liver. It is sexless, but it has kept, to an extraordinary degree, the capacity for cell division; as much

as two-thirds of the liver can be lost and still be replaced by the remaining cells. Why the liver above all
organs? It isn't clear why the liver should be unique;
even though it is a factotum to the body, it is no more
vital than the heart. It does have another property that
is unique among the organs of our body. Its cells have
more than the usual number of chromosomes. Instead
of the normal twenty-three pairs, many cells in the
liver have two, three, or even four times that number:
forty-six, sixty-nine or even ninety-two pairs. Multiple
sets of chromosomes is called polyploidy, and among
animals it is a rare quality; the liver shares its polyploidy with some animals, namely those that creep and
crawl: salamanders, crabs, lobsters, earthworms, and
marine worms. To most other animals, an extra chromosome—or extra sets of chromosomes—are fatal, but
not to these particular groups. Normally, a number of
members of these groups are found in nature to have
extra sets of chromosomes, and they are not harmed
by this extra genetic load. They also can regenerate
whole and complicated parts of their bodies. Salamanders have no trouble replacing an entire leg that has
been severed, complete with muscle and bone.

I think there may be some connection here between
those redundant chromosomes and the ability to recon-
struct an organ or a limb. After all, the liver is the only
organ in our body with such amazing powers of regeneration, and it is also the only organ we have with such
a high proportion of polyploid cells (the lung is polyploid, too, but only slightly). The connection between
extra chromosomes and regeneration holds in the mot-
ley array of creeping creatures already mentioned. It
might be that the presence of those extra genes on
their extra chromosomes allows regeneration. Normally

163

as cells differentiate, become specialized for the tasks they will have to perform, the genes they don't need are turned off, and only the genes they will need when they are part of the mature organ or tissue stay turned on. The extra genes in a polyploid cell might not be turned off, or they may be turned on relatively easily. It's something to think about. Especially when we consider the wonderful (and I mean that as exciting wonder) experiments done with carrots and toads.

These experiments are both famous and infamous. In both experiments, a differentiated, specialized mature cell was taken from each of the organisms: a nutrient-transport cell from the carrot and an intestinal cell from the frog. With the proper treatment (feeding the carrot cell coconut milk and hormones), all of the genes were turned back on. Each of the cells, from the carrot and from the frog, was induced to form an entire new organism; a complete new carrot and a complete new frog. To do this all of the genes had to be turned back on; every one that had been progressively turned off during development. All of the information a cell needs to start back at the beginning as an undifferentiated cell with its whole life ahead of it, remains in the cell.

And we can start all over. We can from a frog's own mature gut cell create (without benefit of fertilization) another full-grown adult frog with all of the appropriate organs, glands, and tissues. The technique is called cloning, intriguing because it solved one of the major problems in the study of development, and macabrely interesting to journalists and the public, who fear (or who would like to instill the fear) that cloning will lead to a sort of Dr. Frankenstein of the cell who creates people all over again from a bit of skin. Some postulate a future in which we are all cloned, in which we are born into this world unborn, replicating as plant lice do without sex, by the transfer of genes into unfertilized eggs which develop to maturity. Virgin birth at last achieved. Maybe. We are considerably different from a carrot and even from a frog. We may one day be able to clone a human being, but these experiments will probably have more real impact upon the way we die than on the way we are born. For the experiments told us what aging and specialization of cells is: a progressive but reversible turning off of unnecessary genes. Because aging originates in our cells; we get old because our cells get old. And we would so very much like to know why it is our cells do not keep us young but betray us to the fate of worms.

We know why in the grand, abstract overview. We have to die for evolution to proceed. To make room. To make progress. Natural selection is monomaniacal about progress. Thomas Malthus told us why, when he catalyzed Darwin's thoughts into the one crystal-clear theory of evolution. In the last century, while everyone postulated the birth of a new golden age through the midwifery of the technology begun in the Industrial Revolution, Malthus was concerned about two things inescapably true: "Food is necessary to the existence of man" and "The passion of the sexes is necessary."

But food increases only arithmetically and people reproduce at a rapid, geometric rate. Against the joy and optimism of the Industrial Revolution, Malthus proposed his "dismal theorem." If the only check on population is starvation and misery (war and disease), then no matter how favorable the environment, or how advanced the technology, the population will grow until it starves or is halted by disease and war. Darwin saw in this the reason why organisms evolve and why they must die. Summarized by a student of mine in calypso, we find:

> The food don't grow as fast as we
> That's Malthusian theory.
> It's really quite unfortunate
> Our growth's geometric.
> Oh, people and animals keep multiplying.
> But most of the extra ones keep dying.

And there are extras—more than the food and space available can provide for. Therefore, it follows as night from the day that, if too many creatures are born, then some must die. Death is a requirement for life. This could be very depressing and it suggests a picture of a death-motivated evolution. An extravagant waste of living creatures. Malthus was preaching, as explicitly as he could in Victorian times, birth control; but nature's birth control seems to take place at the wrong point, and is completely backward. The parents are killed, but after they have reproduced, and after they have produced, as we see from Darwin and Malthus, far too many creatures. This picture was too dismal, for Darwin and apparently for his gentle readers; he had to assure them that it wasn't as brutal as it seemed.

> When we reflect on this struggle, we may console ourselves with the full belief that the war of nature is not incessant, that no fear is felt, that death is generally prompt, and that the vigorous, the healthy and the happy survive and multiply.

Well, perhaps. The sting of death does not seem to be as tender and painless as Darwin would have his readers believe. And evolution can be oriented toward death as well as toward survival, depending upon how you look at it. Each spring thousands of maple seedlings are planted by a single maple tree. Yet, we're not overrun by maple trees because most of these seedlings never make it through the summer or the next winter. Perhaps only one may survive; the others die or are eaten by some browsing animal. Wasteful and extravagant treatment of living things? Not if we think of how many other things have to live on this earth besides maple trees.

Plants are the only intermediaries between the sun and animals, between the animals and starvation. Unless plants live and store this solar energy in edible form, then are eaten or die molding into this earth, everything else would starve. We have to kill them or other animals to obtain the benefits of the plant recycling of the elements unobtainable to us. On the other hand, is genetic experimentation worth all of this death? Well, it would have been a gentler world if we had all stayed amoebas and blue-green algae, but it would have been considerably duller. And we wouldn't have had sex. Or death. Sex and death again.

Sometimes nature makes the relationship painfully clear. Not only are males often smaller than females, their food and growth requirements far less, but they may be killed as soon as they have finished their major chore on earth, exchanging their genes with a female. Spider females are notorious for ensuring that the male does not get a chance to use up any more space or energy than is absolutely necessary. Many kill their mates and eat them after copulation. The males of one particular species of spider, in order to avoid the post-copulatory fate of most male spiders, spins a balloon of web and traps several flies inside it. This balloon is presented to the female, and while she is engaged in examining the gift, the male quickly copulates with her and escapes. This heartless, insensitive destruction of males once they have served their evolutionary purpose is not limited to the arachnids. In certain mice, the male is automatically and naturally disposed of after copulation by a massive rush of hormone from his adrenal glands. Both male and female salmon die after they have spawned. They age and die during their first and last reproductive run upstream. They go from the prime of life through senescence and decay in only two weeks, this rapid condensation of middle and old age triggered by a massive release of the same adrenal hormone. Procreation is done, and so are their lives. There has to be room and food. Nothing extraneous is allowed to survive for long in nature. Survival is an odds-against proposition in this living world. It requires all of that death—salmon and maple trees, mice and men—to ensure survival.

Although we do not necessarily treat the males of our species so unkindly, we are caught in this same web, the web of required mortality. Death supplies life. Plants and animals are food for human beings and other carnivores, and we, in turn, are food for worms, fungus, and bacteria. Or at least we were before we decided against returning to the bosom of our mother earth, and chose to seal ourselves in airtight concrete and metal instead. We have refused to participate in this particular universal activity. We would also like to be able to refuse to participate in dying, to avoid the ashes-to-ashes and dust-to-dust judgment. We would like to escape completely, but failing that, we want to delay the final execution. Surgery and medicine have fed our illusions, helped us to believe what we wanted to believe, not what Darwin told us. We are, he said, not celestial creatures, one step below the angels, but very earthy mortal apes. Mortal. Our ego helped us fend off the arguments for Darwin's ideas for a long time. Chimpanzees may have a certain disconcerting, hairy pseudohuman charm, but only when they are swinging from something other than the branches of our family tree.

We were also reluctant to reevaluate the powers that rule our lives, as Darwin forced us to do. We had once devoted a great deal of attention to attempts to know and worship whatever powers, whatever gods there were. We were creatures of their special creation, and so it behooved us to find out as much about our creator or creators as we could. We are turning away more and more, from those particular tasks. The properties of the gods no longer hold interest for us as we become increasingly sure that it is ourselves who control our destinies. As we grow in that particular self-assurance, how galling it must be to recognize that the last, the final and finishing touch of our destinies we cannot control at all. We rebel against this, and perhaps our rebellion is accentuated by a sense of desperation. For as we abandoned the idea of a real and actual, if not knowable, god or gods, we have had to abandon the idea of a hereafter, or at least revise it drastically. It all becomes too frightening a thought, if we combine the knowledge that we are mortal with the suspicion that there may be nothing after that, nowhere to go. We find ourselves increasingly reluctant to chance that final leavetaking, and increasingly interested in aging and death, since it seems that we might be able to do something about them. To circumvent nature's plan for us. It's only fair. After all, nature is very hostile to senior citizens. It's only fair that we refuse to let death come when it may come. After all, we have to show who is in control here, don't we?

At the moment, as always, our cells are in control. They are at the center of death as they are at the center of life. In our eyes, however, they let us down. They make mistakes. And death is the final large mistake. Our cells are no longer working as they did, for the good of the organism. It is because our code of life breaks down that the machine stops, some thought.

In our genetic translation machinery there are too many places where mistakes can happen, and since the machinery can't detect its own mistakes, these errors accumulate. In the genes, in the transcription of the genes into messenger. The translation into proteins can get sloppy, too; the structures and molecules worn out with age as the parts of any much-used machine would be worn. This theory made sense, but investigation proved otherwise. We get old and die because our cells no longer divide when they were past a certain age, others said. And the intriguing Hayflick number was discovered. Hayflick found that human blast cells, which are among the few kinds of cells in the body that keep their immature capacity to divide, divide only fifty times (plus or minus ten times). No matter if they were left alone, interrupted, transferred to different culture flasks, each cell would divide just fifty times. No matter what Hayflick did to these cells, they seemed to keep a memory of how many times they had divided—even if they had been frozen and stored in suspended animation for six years. If the frozen cells were thawed they would continue dividing, but never more than fifty divisions in all. These experiments gave us the magical Hayflick number, a number that cells know and obey. The blast cells always kept to their schedule. And this schedule may be part of the death that is programmed in our genes. It could very well be that each cell, in its whole lifetime is not allowed more than fifty divisions, and once they have reached their Hayflick number they have reached the end of their cellular lives. Even though each part of our body dies at a different rate, the same organ in different persons dies at a constant rate.

If we compare the amount of function lost between the ages of thirty (when aging starts) and seventy-five (when death usually stops the aging process) in different organs, we find that almost everyone's brain weight has decreased by 8 percent. The rate of nerve conduction has slowed by 10 percent. The blood supply to the brain has been reduced by 20 percent. Our kidneys do their filtering 31 percent more slowly. Our lungs have lost 44 percent of their capacity. Our heart pumps 30 percent less blood than it used to. We have lost 37 percent of the axons in our spinal nerves. We have only 36 percent of our taste buds left. A sad inventory, and the Hayflick number may be responsible; cells of different organs have different rates of division, but all are ruled by the Hayflick number. All animals that grow as long as they live, like fish, age much more slowly than mammals and birds, which stop growing at maturity. Those plants that reach prodigious ages, like the sequoia, are able to stay alive that long because they form new tissues and cells year after year, while the cells of the old tissues die off. A giant redwood is 4,000 years old, perhaps, but it contains no living cells that are older than a few years. Trees and other woody plants have a tissue—the meristem—that remains perpetually youthful, constantly dividing to form brand new cells. The aging and death of the individual cells in an organism and the aging and death

171

of the entire organism can be two entirely different things. And it is the aging of the cells—whether or not they are allowed to get old—that is the most important.

And it is getting old that we are frightened of perhaps more than the actual leavetaking. After all, about two-thirds of the American public have contemplated suicide. But no one wants to get old, to experience the progressive, inescapable breakdown of the beautiful, indescribable functions of the once-magnificent machine of our body. Aging might be more unpleasant to contemplate than the time when the machine finally stops. We know we can live a little longer by following some fairly simple rules of diet, habits, exercise, but all this does is help more of us reach our threescore and ten. It just gives us more of a chance of succumbing to some disease. One predictable and terrible breakdown is that of our immune system, leaving us helpless and susceptible to bacteria, virus, fungi, and cancer. Even as our vision fades, so does the recognition ability of the immune system. Self is mistaken for nonself, and the body starts attacking its own tissues, and among autoimmune diseases, we get arthritis, the painful characteristic affliction of age.

There is a general fading of the immune system in general; older people produce fewer T cells, the cells in our immune system that specifically recognize and combat virus and fungi and reject tumors. The T cells are the cells in our immune system that have spent some time in our thymus gland, a really extraordinary gland with a peculiar and indicative history. The immune system is the only one of our systems that begins working only after birth, and the thymus is responsible for the original construction of our immune system. The thymus is large at birth and, after it has supervised the installation of the immune system, it gradually disappears until at adolescence it is not more than a few fibers, and almost impossible to find in an adult. It is this gland of our youth and its early disappearance that may mark our first step toward aging and death. With fewer T cells, we then are extremely vulnerable to the second greatest killer, cancer. A breakdown of the immune system marks aging and aging in mice can be retarded merely by transplanting a new thymus gland into them. It can also be retarded by transplants of bone marrow, which is the source, in humans as well as in mice, of the B cells, the important antibody-producing cells of the immune system. We are not surprised by this betrayal by our immune system; it is just another example of how our cells fail us and make us old and frail.

We may not be surprised either by the fact that certain cells in our brain have been found predictably to fail with age; the cause and treatment of Parkinson's disease, whose dreadful palsy and weakness are like premature aging and senility, may give us important clues about how our brain ages. We know that we can treat the symptoms of Parkinson's disease successfully with massive doses of the drug L-dopa. L-dopa is a precursor in the body to the synthesis of dopamine, a molecule responsible for transmitting neural messages in certain parts of the brain. Dopamine carries messages between the hypothalamus and the pituitary gland. Our pituitary gland is located just north of our ears and midway between, tucked directly underneath the brain. The pituitary controls our growth, reproduction and metabolism, which of course manages the physiology of our youth. Two parts of the brain itself are linked to youth: the thalamus controls muscle movement, and the hypothalamus directs the operation of the pituitary that releases the hormones, the chemical essences of our youth, giving us growth, rapid metabolism and reproductive talents. One of the substances that transmits messages back and forth among the thalamus, the hypothalamus and the pituitary is dopamine; and in aging mice there is very little dopamine, especially in the thalamus and in the part of the hypothalamus that directly controls the pituitary. Although dopamine may be necessary to keep the brains of mice youthful, and extends the lives of Parkinson patients by about 10 percent, it is neither a completely successful treatment for Parkinson's nor a good candidate as a rejuvenating treatment. The massive doses used to treat Parkinson's disease cause severe schizophrenic seizures.

We become more susceptible to disease.

Our brains don't work as well as they did.

We lose the lithe flexibility of youth.

We know a great deal about how our connective tissues age, our tendons and cartilage. We get rigid and less flexible with age because the tissue that holds our bones together and helps them move gets rigid and less flexible. We lose the loose grace of youth because the ground substance of our connective tissues becomes thicker, and the elastic fibers in this become thicker and less elastic as well. Strangely, we cannot say where aging begins here, for the aging of connective tissue is purely and simply a direct continuation of the processes that began in the embryo at the initial formation of connective tissue. But these processes never stop; they continue without halt. The fibers and their ground substance are brought to their appropriate rigidity, but still the building continues. And too soon we are made

174

to creak about. It is truly peculiar, a system of development that cannot be stopped once it is set in motion. But these are all just symptoms of aging. They tell us what happens when we get old. But it is merely a list of physical flaws; scars instead of cells are used to repair damage, molecules are not produced, our immune system fails, our brain does not work as well as it did, we move with less ease than we did once. Waste products may build up inside the cell and interfere with its function, and there is evidence that inactive enzymes accumulate.

This is not why we age; this is how we age. The real essence of the question is why our cells let us down. Ah, that. Well, that, at present, is not known. There is quite good evidence though that our cells let us down because they are not getting instructions from the genes. Cross-links form between the strands of that spiral of life, the DNA double helix. The two strands cannot separate and the genes cannot be read. The cell does not know what to do, so it does nothing or it does something wrong.

An enzyme has been discovered, we are told, as we are told many hopeful things. And this enzyme can break these cross-links and free the genes, doubling our life span. It could be a chemical Shangri-la, this enzyme, keeping us young, not forever but for extra-life span at least. Whether the enzyme can be used to prevent human aging is not known. How we can get it into our cells is not known. (After all our cells may be fiercely dedicated to dying at the preordained time.) Whether we will want to use the enzyme even if it works is not known. We all weep, as in Gerard Manley Hopkins's poem, Margaret did, "over goldengrove unleaving" in the fall—most of us without a poet to ennoble our very human grief. For us, as for Margaret, it isn't really the falling leaves that sadden us so.

> It is the blight man was born for
> It is Margaret you mourn for.

But there might be a surprising number of us who sincerely mourn the passing of our lives, and the final fate that faces us, yet who will, after taking a good look at our species and at our world, refuse an injection of that prolonging enzyme.

For the hereafter is a purely speculative venture, and since atoms don't die and seem to have some self-patterning capabilities—perhaps even memory—dying might turn out to be a far more attractive alternative than extending our lives through enzymatic tinkering. The atomic memories of us, the knowledge of our innermost essence might be a lovelier monument than any of marble, gilded or not. Lovelier than a taste of immortality. For our atoms might carry our personal memorabilia with them into other forms of life, if only we will let our atoms go. So we should consider well how we treat them, these atoms of ours; be careful of their sensibilities so their memories of us will not be bitter ones—that they do not remember being locked away hermetically and unnaturally, as huddled masses yearning to breathe free . . . that the atoms of our DNA do not remember and resent dilettante dabbling with their evolutionarily imposed cross-links.

Have care of your atoms, for such is the stuff that dreams are made of—the yearning dreams of our immortality.

ABOUT THE AUTHOR

L. L. LARISON CUDMORE received her Ph.D.
in zoology from Yale and studied as a research
Fellow at Harvard University for two years. She
has taught at the University of Massachusetts
and at Boston University, and is among the leading theorists and experimentalists in cell and
molecular biology.

QUADRANGLE BOOKS

SCIENCE

W. E. LeGros Clark	THE ANTECEDENTS OF MAN	4.95
Robert Cooke	IMPROVING ON NATURE	3.45
L. L. Larison Cudmore	THE CENTER OF LIFE	2.95
Claude A. Frazier	COPING WITH FOOD ALLERGY	3.95
Arthur Vineberg, M.D.	HOW TO LIVE WITH YOUR HEART	2.95

PSYCHOLOGY

Sol Gordon with Roger Conant	YOU	6.95
Janet and Paul Gotkin	TOO MUCH ANGER, TOO MANY TEARS	4.95
Florence R. Miale and Michael Selzer	THE NUREMBERG MIND	3.95
Patricia O'Brien	THE WOMAN ALONE	3.95
Edward Rosenfeld	THE BOOK OF HIGHS	4.95
Charles W. Socarides	BEYOND SEXUAL FREEDOM	2.95
Paul Tibbetts	PERCEPTION	4.95

GENERAL & REFERENCE

John Bailey	INTENT ON LAUGHTER	2.95
Ada Louise Huxtable	KICKED A BUILDING LATELY?	5.95
Mike Jahn	ROCK	4.95
Arnold and Connie Krochmal	A GUIDE TO THE MEDICINAL PLANTS OF THE UNITED STATES	4.95
Muriel Lederer	THE GUIDE TO CAREER EDUCATION	6.95
Miriam Makeba	THE WORLD OF AFRICAN SONG	3.95
Nancy M. Page and Richard E. Weaver, Jr.	WILD PLANTS IN THE CITY	3.95
Dr. John J. Reynolds and Dr. Thomas D. Houchin	A DIRECTORY FOR SPANISH-SPEAKING NEW YORK	3.95
Richard F. Shepard	GOING OUT IN NEW YORK	4.50